ASCE Manuals and Reports on Engineering Practice No. 136

Concrete Foundations for Turbine Generators

Analysis, Design, and Construction

Task Committee on Turbine Generator Foundations
Edited by Hongchun Liu, P.E.

Sponsored by the
Energy Division of the
American Society of Civil Engineers

ASCE AMERICAN SOCIETY OF CIVIL ENGINEERS

Published by the American Society of Civil Engineers

Library of Congress Cataloging-in-Publication Data

Names: Liu, Hongchun, editor. | American Society of Civil Engineers. Task Committee on Turbine Generator Foundations.
Title: Concrete foundations for turbine generators: analysis, design, and construction / Task Committee on Turbine Generator Foundations; edited by Hongchun Liu, P.E.
Description: Reston, Virginia: American Society of Civil Engineers, [2018] | "Sponsored by the Energy Division of the American Society of Civil Engineers." | Includes bibliographical references and index.
Identifiers: LCCN 2017053754 | ISBN 9780784414927 (soft cover: alk. paper) | ISBN 9780784481097 (pdf) | ISBN 9780784481103 (ePub)
Subjects: LCSH: Turbogenerators—Foundations—Design and construction. | Steam-turbines—Foundations—Design and construction. | Concrete footings—Design and construction.
Classification: LCC TK2458.F68 C66 2018 | DDC 621.31/3—dc23 LC record available at https://lccn.loc.gov/2017053754

Published by American Society of Civil Engineers
1801 Alexander Bell Drive
Reston, Virginia 20191-4382
www.asce.org/bookstore | ascelibrary.org

Errata: Errata, if any, can be found at https://doi.org/10.1061/9780784414927.

ISBN 978-0-7844-1492-7 (print)
ISBN 978-0-7844-8109-7 (PDF)
ISBN 978-0-7844-8110-3 (ePub)
Manufactured in the United States of America.

25 24 23 22 21 20 19 18 1 2 3 4 5

MANUALS AND REPORTS ON ENGINEERING PRACTICE

(As developed by the ASCE Technical Procedures Committee, July 1930, and revised March 1935, February 1962, and April 1982)

A manual or report in this series consists of an orderly presentation of facts on a particular subject, supplemented by an analysis of limitations and applications of these facts. It contains information useful to the average engineer in his or her everyday work, rather than findings that may be useful only occasionally or rarely. It is not in any sense a "standard," however; nor is it so elementary or so conclusive as to provide a "rule of thumb" for nonengineers.

Furthermore, material in this series, in distinction from a paper (which expresses only one person's observations or opinions), is the work of a committee or group selected to assemble and express information on a specific topic. As often as practicable the committee is under the direction of one or more of the Technical Divisions and Councils, and the product evolved has been subjected to review by the Executive Committee of the Division or Council. As a step in the process of this review, proposed manuscripts are often brought before the members of the Technical Divisions and Councils for comment, which may serve as the basis for improvement. When published, each work shows the names of the committees by which it was compiled and indicates clearly the several processes through which it has passed in review, so that its merit may be definitely understood.

In February 1962 (and revised in April 1982), the Board of Direction voted to establish a series titled "Manuals and Reports on Engineering Practice," to include the Manuals published and authorized to date, future Manuals of Professional Practice, and Reports on Engineering Practice. All such Manual or Report material of the Society would have been refereed in a manner approved by the Board Committee on Publications and would be bound, with applicable discussion, in books similar to past Manuals. Numbering would be consecutive and would be a continuation of present Manual numbers. In some cases of joint committee reports, bypassing of Journal publications may be authorized.

A list of available Manuals of Practice can be found at http://www.asce.org/bookstore.

CONTENTS

PREFACE

In 1987, the Task Committee on Turbine Foundations of the Fossil Power Committee and the Nuclear Energy Committee of the Energy Division of the American Society of Civil Engineers (ASCE) published the first guide for the design of large steam turbine generator foundations.

Since the guide's publication, there have been significant changes in design codes, advances in computer analysis, and increased industry pressure to refine foundation designs. A new task committee was formed in April 2014, under the leadership of the ASCE Energy Division chaired by J. G. (Greg) Soules, to address the wide variation of analysis, design, and construction practice that have been applied throughout the industry for turbine generator foundation designs over the past 30 years in the United States.

Designing a turbine generator foundation requires not only advanced technical expertise in structural and soil dynamics, but also close collaboration with machine manufacturers and other disciplines (mechanical, electrical, and plant design). Different manufacturer requirements, and code provisions that are subjected to interpretations when applied to such foundations, present additional challenges to the design engineers.

This document was prepared by the task committee to provide practical guidance in the analysis, design, and construction of turbine generator foundations. It is a result of significant collaborative efforts from all committee members, as well as contributions from other industry experts and professionals.

The intended audience for this document includes structural design engineers, operating company personnel responsible for establishing structural design criteria and construction standards, and local building officials.

For more than three years, a number of key individuals dedicated significant amounts of time and effort to formulating, originating, and reviewing in detail, either the whole document or specific sections thereof. Those members are identified in the Acknowledgments.

ACKNOWLEDGMENTS

Task Committee on Turbine Generator Foundations

Chair
Hongchun Liu, P.E.
Bechtel Infrastructure & Power Corp.

Vice-Chair
Abbas Mokhtar-zadeh, Ph.D., P.E., PMP
Westinghouse

Members

Henning Almstedt	Siemens AG Power and Gas Division
Jaya Brata Bose, P.E.	Bechtel Infrastructure & Power Corp.
Carlos Coronado, Ph.D., P.E.	Bechtel Oil, Gas & Chemicals Corp.
Julie Drozd Gennaro, MSCE	AECOM
Jaime Guamán Cabrera, MSc.	
Kyle Harris, P.E.	Kiewit
Ping (Philip) Jiang, P.E.	WorleyParsons Group Inc.
David Kerins, P.E.	ExxonMobil Research and Engineering Company
Hoan-Kee Kim, Ph.D., P.E.	ENERCON
John (Zhong) Liu, Ph.D., P.E., S.E.	Kiewit
Gerald Miller	GE Power
Hesham El Naggar, Ph.D., P.E.	Western University, Canada
Peter Nawrotzki, Ph.D.	GERB Vibration Control Systems
Daniel Predmore	GE Power
Stephan De Roo	Siemens AG Power and Gas Division
Vincenzo Sofia, P.E.	GE Power
Mansour Tabatabaie, Ph.D., P.E.	MTR & Associates, Inc.
Frederick Tajirian, Ph.D., P.E.	Chevron Energy Technology Company
Xuan Wang, Ph.D., P.E.	Sargent & Lundy LLC.
F. Alan Wiley, P.E.	Black & Veatch Corp.
Derek Winkler, P.E.	AECOM

Contributors

The following individuals are recognized for their reviews and other contributions in the preparation of this book:

Siddhartha Bhattacharya, P.E.	Bechtel India Pvt. Ltd.
Rami Said, P.E.	Siemens Energy Inc.
Victor Salcedo	GERB Vibration Control Systems
Yasser Salem, Ph.D., S.E.	California State Polytechnic University
Shen Wang, Ph.D., P.E.	TerraPower LLC.

Blue Ribbon Review Panel

A Blue Ribbon Review Panel, consisting of seven independent reviewers and formed per the ASCE requirements on publishing Manuals and Reports on Engineering Practice (MOPs), has reviewed this book.

William Bounds, P.E.	Fluor Corp.
Mukti Das, Ph.D., P.E., F.ASCE	MLD Engineering
Shu-Jin Fang, Ph.D., P.E., S.E.	Sargent & Lundy LLC.
George M. Ghaly, M.Sc., P.E.	Westinghouse
Luis Moreschi, Ph.D., P.E.	Bechtel National, Security & Environmental Corp.
Farhang Ostadan, Ph.D., P.E.	Bechtel National, Security & Environmental Corp.
Norman Rennalls, P.E.	Shell Corp.

The task committee would like to thank the following ASCE 7 committee members for their contributions:

Robert E. Bachman, P.E.
Thomas F. Heausler, P.E., S.E.
Dominic J. Kelly, P.E., S.E.
Sanjeev R. Malushte, Ph.D., P.E., S.E.

The task committee also thanks Robert (Bruce) Dongarra for drafting numerous sketches for this work.

Finally, the task committee thanks the ASCE Energy Division Executive Committee, Jay Snyder and Barbara Whitten, technical managers at ASCE, and Betsy Kulamer, Director of Books at ASCE, for their continuous support during the preparation of this book.

CHAPTER 1
INTRODUCTION

1.1 BACKGROUND

In 1987, the Task Committee on Turbine Foundations of the Fossil Power Committee and the Nuclear Energy Committee of the Energy Division of the American Society of Civil Engineers published the first guide for the design of large steam turbine generator foundations (ASCE Task Committee on Turbine Foundations 1987). This initial report attempted to organize into a single document all the criteria necessary to analyze and design large steam turbine-generator (STG) foundations.

Since the guide's publication, there have been significant changes in design codes, advances in computer analysis, and increased industry pressure to refine the design of turbine generator (TG) foundations. A new ASCE task committee on TG foundations was formed within the energy division of ASCE in 2014 to review the industry changes and their impact on the design of these large concrete structures.

The new ASCE Task Committee on Turbine Foundations has revised the 1987 guide to address current industry practices and to expand content to include information on these topics:

- Design of combustion turbine generator (CTG) foundations, in addition to steam turbine generator (STG) foundations.
- Development of finite element (FE) analytical models.
- Development of dynamic stiffness and damping for dynamic analyses of soil- and pile-supported STG and CTG foundations.
- Design of embedded items.
- Construction considerations specific to TG foundations.

The Steel Foundations chapter of the 1987 guide is not in the scope of this book, because a vast majority of TG foundations are constructed using reinforced concrete.

The new task committee members are representatives from leading engineering firms involved in the design of power plants, major turbine generator manufacturers, manufacturers of vibration isolation systems, and universities. The diversity of this committee is intended to provide expertise on all aspects of TG foundation design.

1.2 PURPOSE

This book is intended as a state of practice guide for practicing engineers and designers responsible for the analysis and design of gas and steam turbine generator foundations. Recommendations for the design, analysis, construction, and other aspects of TG foundations are provided as deemed adequate by the committee. The responsibility for satisfying the manufacturer's criteria for a particular installation is left to the responsible engineers.

This report is not intended to supersede any manufacturer's criteria.

1.3 SCOPE AND LIMITATIONS

The scope of this book is limited to the design of the following concrete TG foundations:

- Block Foundation—pedestals, piers, and blocks supported on a common basemat.
- Elevated Space-Frame Pedestal Foundation—tabletop and columns supported on a basemat.
- Vibration Isolated Foundation—equipment or structural elements supported on vibration isolation systems.

This manual of practice does not address steel foundations.

This book is organized into eleven chapters with three appendixes. A brief summary of each chapter is provided as follows.

Chapter 2: Turbine Generator Equipment. In Chapter 2, the main components of TG systems are described, including combustion turbines (CT), steam turbines (ST), condensers, and generators. Typical TG system layout configurations are discussed and illustrated. An overview of foundation design, installation, and operation is also presented.

Chapter 3: Preliminary Turbine Generator Foundation Layout and Sizing. The objective of Chapter 3 is to provide guidance for preparing the overall foundation layout and initial sizing of the foundation's structural elements. Guidance is also given on the data and documents required to develop the preliminary design and for performing the analyses described in subsequent chapters.

Chapter 4: Foundation Loads and Load Combinations. In Chapter 4, the load cases that must be considered for the analysis and design of TG foundations are discussed with information on development of these loads and how to apply them. Recommendations for developing the load combinations for the foundation strength design are also provided.

Chapter 5: Modeling of the Soil and Pile Response to Dynamic Loads. One major challenge in TG foundation design is to ensure that the foundation's dynamic responses do not impose adverse impacts on the normal operation of the machine. The stiffness and damping (i.e., dynamic impedance) of the combined foundation and soil or pile system plays an important role in this regard. From a structural engineer's perspective, Chapter 5 provides guidance on how to appropriately consider dynamic impedances of soil or pile when performing the dynamic analysis of the TG foundation. Practical discussions on its applicability and limitations are also provided.

Chapter 6: Finite Element Modeling. The objective of Chapter 6 is to provide technical guidance to develop an appropriate FE model that can sufficiently capture the structural characteristics of a TG foundation. Various types of elements used in FE models are discussed, along with other modeling considerations for TG foundations.

Chapter 7: Serviceability Analysis and Acceptance Criteria. Chapter 7 describes typical serviceability analysis and acceptance criteria required by equipment manufacturers for the design of TG foundations. Information on both dynamic and static serviceability requirements is provided, including applicable loads, load combinations, and acceptance criteria.

Chapter 8: Strength and Stability Design. Chapter 8 covers strength design of TG foundations, including seismic load and ductile design considerations, reinforcement requirements unique to TG foundations, guidance for pile capacity, and soil-bearing pressure checks. Special design considerations and detailing requirements are also provided.

Chapter 9: Embedded Items. Chapter 9 describes the function, arrangement, and design of various embedded items used in TG foundations. These include anchor bolts, jacking posts, shear lugs (keys), embedded plates, and other items used to support electrical and piping components. The selection of suitable grout material for TG foundations is also discussed.

Chapter 10: Vibration Isolated Foundations. Chapter 10 describes the design properties of the spring elements and viscous dampers constituting the vibration isolation system (VIS), which, if properly used, allow the TG machine to operate independently from the substructure and dynamic subsoil conditions. This chapter provides guidance in the design of TG foundations supported on VIS. Typical layouts of VIS are given and advantages of using VIS are also discussed.

Chapter 11: Construction Considerations. Chapter 11 provides guidance on addressing construction issues unique to TG foundations: construction joints to allow multiple concrete pours; headed reinforcement to reduce rebar congestion; suitable concrete mix to eliminate placement issues; rebar modularization for ease of installation; and other considerations. Industry best practices are highlighted and innovative ideas are explored, which may help reduce costs and accelerate construction schedules.

Appendixes A, B, and C: Supplemental Information on Foundation Dynamic Impedances and Calculation Examples. These appendixes provide valuable supplemental information to the contents in Chapter 5 on both soil and pile dynamic impedances. Calculation examples are provided to illustrate the process of determining these important design parameters.

1.4 ABBREVIATIONS AND ACRONYMS

CT	Combustion Turbine
CTG	Combustion Turbine Generator, a.k.a. Gas Turbine Generator (GTG)
DAF	Dynamic Amplification Factor
FE	Finite Element
FEA	Finite Element Analysis
FEM	Finite Element Model
GA	General Arrangement
GTG	Gas Turbine Generator, a.k.a. Combustion Turbine Generator (CTG)
HP	High-Pressure
HRSG	Heat Recovery Steam Generator
IP	Intermediate-Pressure
LP	Low-Pressure
MPF	Mass Participation Factor

RPM	Revolutions per Minute
SDC	Seismic Design Category
SSI	Soil-Structure Interaction
ST	Steam Turbine
STG	Steam Turbine Generator
TG	Turbine Generator
VIS	Vibration Isolation System

1.5 FOUNDATION TERMS

It is recognized that various parts of a foundation are referred to by many different names. The following nomenclature has been adopted for this manual of practice:

Basemat (or Mat Foundation)—the part of the TG concrete foundation supported on soil, rock, or piles, consisting of a continuous concrete slab extending in both directions, reinforced and supporting an array of columns, piers, and/or walls. It is also called a pile cap when the basemat is supported by a group of piles.

Block Foundation—a concrete foundation that has individual pedestals, piers, and blocks, sharing a common basemat.

Column—a vertical structural element, used primarily to support axial compressive loads, but can also resist moment, shear, or torsion, with a height-to-least-lateral-dimension ratio exceeding 3.0.

Critical Speed—the theoretical angular velocity that excites the natural frequency of a rotating object, such as a shaft, propeller, leadscrew, or gear.

Dynamic Stiffness—in structure dynamics, it refers to the ratio of the change of force to the change of displacement under dynamic conditions, or the ratio of input force to output displacement. In soil dynamics, it refers to the real part of the dynamic impedance formulated by a spring-dashpot analog.

Elevated Space-Frame Pedestal Foundation—a concrete space-frame structure that has a tabletop supported by vertical structural elements such as columns, pedestals, and walls, which are supported on a common basemat.

Longitudinal Girders—large horizontal beams that serve as the main structural members supporting the TG set, parallel to the axis of the machine.

Machine Foundation—a system of structural elements, including tabletop, columns, pedestals, walls, and basemat, which transmits loads from the machine to the supporting soil.

Pedestal—a vertical structural element with a height-to-least-lateral-dimension ratio not exceeding 3.0. Used primarily to support axial compressive loads.

Rebar—reinforcement bars.

Soil—a generic term for unconsolidated natural surface material above bedrock.

Superstructure—the part of the foundation above the basemat, consisting of the pedestals, columns, walls, and tabletop.

Tabletop—the uppermost portion of the foundation, consisting of longitudinal girders, transverse beams, and slabs.

Transverse Beams—structural members subjected primarily to flexure that support the turbine-generator set, transverse to the axis of the machine.

Vibration Isolated Machine Foundation—a foundation that has the equipment or structural elements supported on a vibration isolation system.

Wall—a vertical structural element used to primarily transfer axial compressive load and lateral loads to the basemat.

REFERENCES

ASCE Task Committee on Turbine Foundations. (1987). *Design of large steam turbine-generator foundations*, February 1987.

ACI (American Concrete Institute). (2011). "Foundations for dynamic equipment." *ACI 351.3R-04*, Farmington Hills, MI. Reapproved 2011.

CHAPTER 2
TURBINE GENERATOR EQUIPMENT

2.1 INTRODUCTION

In a modern power plant, both STG and CTG systems generate electricity by converting thermal energy to other forms of energy. The fundamental difference between the two systems is how the energy used to rotate the generator shaft is produced. An STG system utilizes high-temperature pressurized steam that expands through turbine section(s) to do the mechanical work, whereas a CTG system utilizes combustion of air and natural gas fuel to generate a high-temperature pressurized gas flow that expands to do the same.

STG and CTG machines are arranged in a variety of configurations. All TG sets include at least one turbine and one electrical generator connected by a coupled rotor shaft, among other additional components.

2.2 MAIN COMPONENTS

2.2.1 Combustion Turbines

Combustion turbines are typically comprised of a number of sections: inlet, compressor, combustion turbine, and exhaust. The turbine section is where the energy of the compressed, heated air and combustion of gas is extracted.

2.2.2 Steam Turbines

Steam turbines are typically comprised of one or more turbine sections, classified as high-pressure (HP), intermediate-pressure (IP), and

Fig. 2-1. A turbine rotor without casing
Source: Courtesy of Siemens AG; reproduced with permission.

low-pressure (LP), in which the steam expands from high to low pressure with the low-pressure turbine eventually exhausting to one or more condensers. In some multisection configurations, after the steam is expanded in one section, it is routed externally for reheating and moisture removal before returning to the next section for further expansion.

A reliable heat source is required to produce a sufficient quantity of inlet steam for steam turbines. In combined-cycle configurations, exhaust heat from the combustion turbine is used for steam production using the heat recovery steam generator (HRSG). In geothermal steam turbines, the steam is taken directly from underground sources. Other heat sources for steam turbines are coal combustion, solar fields, and nuclear reactors.

Turbine rotors (comprised of a shaft and blades, as shown in Fig. 2-1) within the turbine sections are supported on rotor bearings.

2.2.3 Condensers

In an STG configuration, a condenser is usually required to cool the exhaust steam, which is then returned to the steam supply system in the form of condensate liquid water. A vacuum is formed inside the condenser from the pressure reduction, concurrent with temperature reduction occurring as the steam changes from a vapor to a liquid state. A condenser is typically arranged in-line with the turbine centerline or normal to the turbine centerline and can be located above, below, or off to the side of the turbine.

Depending on how the steam is cooled, condensers are categorized as one of two types: water-cooled or air-cooled. Water-cooled condensers cool by circulating cooling water through thin-wall tubes within the condenser shell to enhance heat transfer without contacting or mixing with the condensate. Air-cooled condensers cool by circulating ambient air around finned tubes containing the condensate. Water-cooled condensers are smaller than air-cooled condensers from the higher cooling capacity of water over air, but require a reliable source of cooling water. In locations where cooling water is not available, air-cooled condensers are used.

2.2.4 Generators

Generators are comprised of a stator, a rotor, and an exciter. The stator is a circular frame with armature coils arranged in slots in its inner circumference. The rotor has field coils wound on its external circumference. The exciter is a device that energizes the field coils on the generator rotor with direct electrical current. Exciters may be mounted directly on the foundation or on the generator stator frame. Some exciters include an additional rotor shaft coupled to the generator rotor.

During operation, the coupled TG rotor is driven to rotate by the expansion of the working flow in the turbine sections. The energized, rotating field coils on the generator rotor produce an electromagnetic field that induces an alternating electric current in the generator stator armature.

2.3 LAYOUT CONFIGURATIONS

TGs are classified as having either a single-shaft or a multi-shaft configuration. In single-shaft configurations, turbine sections and the generator are arranged in series along a single rotor shaft. In multi-shaft configurations, turbine sections are arranged on two or more separate shafts, with an independent generator for each shaft. The separate shafts rotate independently. The rotors within a given shaft are connected to each other by couplings.

TG rotors are supported by bearings in bearing support structures. These support structures can be integral with the turbine or generator stator casings or separate bearing housings. Some designs use two bearings per rotor, one at each end, while other designs share single bearings between major components.

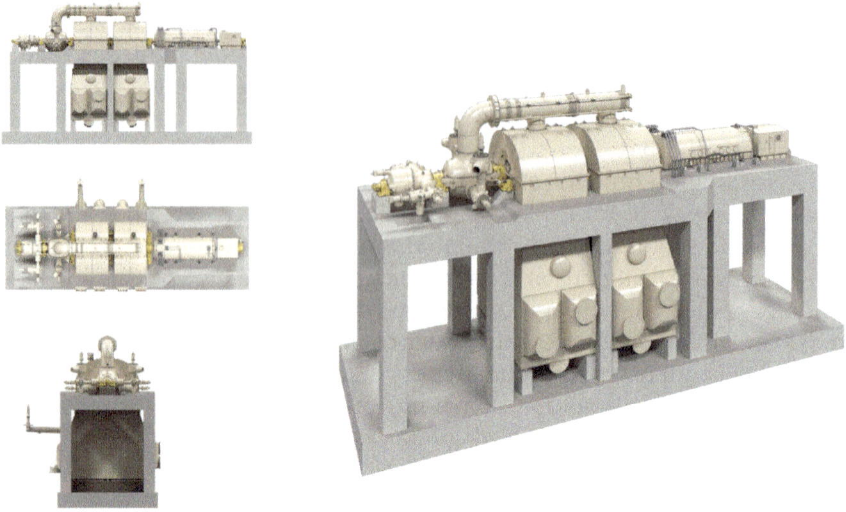

Fig. 2-2. STG arrangement: HP-IP-LP-LP with bottom exhaust
Source: Courtesy of Siemens AG; reproduced with permission.

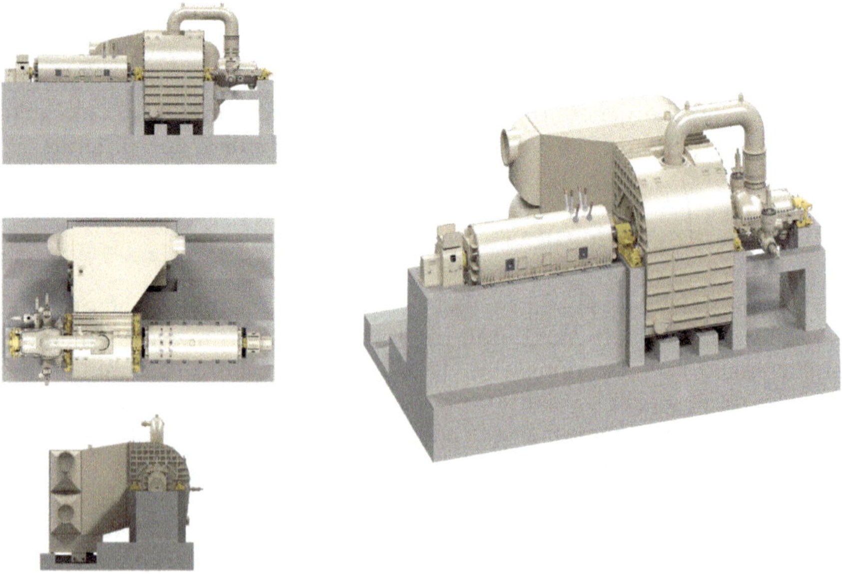

Fig. 2-3. STG arrangement: HP/IP-LP with side exhaust
Source: Courtesy of Siemens AG; reproduced with permission.

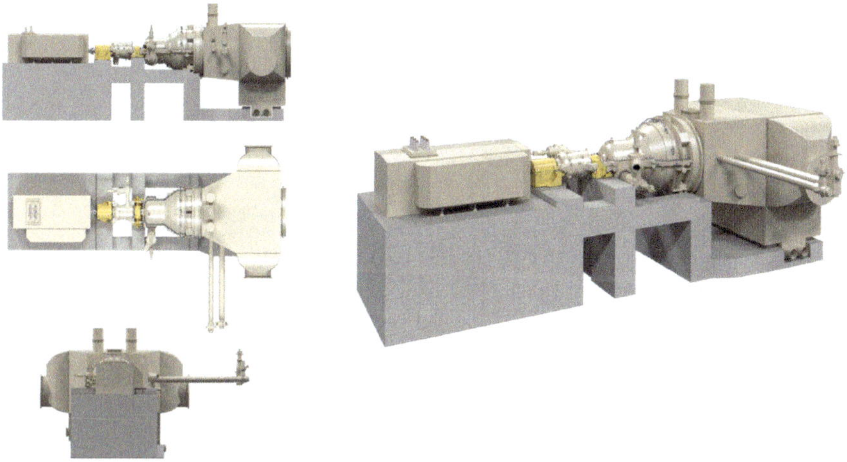

Fig. 2-4. STG arrangement: HP-IP/LP with axial exhaust
Source: Courtesy of Siemens AG; reproduced with permission.

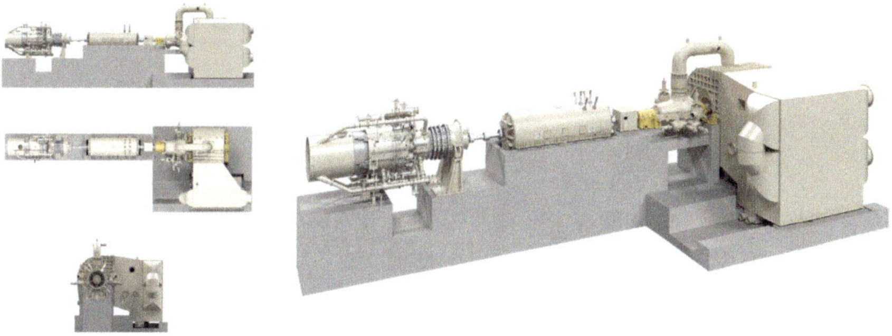

Fig. 2-5. STG arrangement: HP/IP-LP in single shaft with side exhaust
Source: Courtesy of Siemens AG; reproduced with permission.

An individual turbine section may be classified as axial-flow or opposed-flow. In axial-flow sections, the working flow expands in one axial direction. In opposed-flow sections, the working flow enters at or near the center of the section and expands in opposite axial directions. The staging in an opposed-flow section may or may not be symmetrical.

As a result of customized designs and applications, there are a large variety of TG layout arrangements in modern power plants. Figs. 2-2 to 2-7 show several sample layout arrangements for STG and CTG equipment.

For bottom or down exhaust condenser arrangements, there are three major methods of support for the condenser (Fig. 2-8).

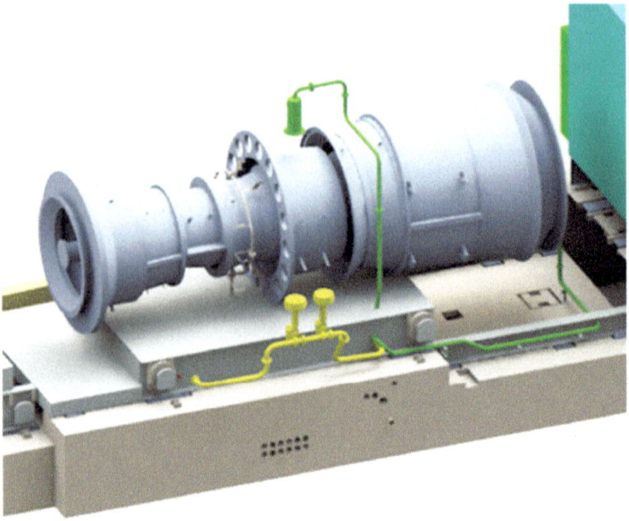

Fig. 2-6. A combustion turbine on a mat foundation
Source: Courtesy and copyright of General Electric Company; all rights reserved.

Fig. 2-7. A combustion turbine on a pedestal foundation
Source: Courtesy and copyright of General Electric Company; all rights reserved.

1. The bottom of the condenser is mounted on rigid supports, and an expansion joint is placed between the condenser and the turbine exhaust nozzle to relieve thermal forces and variations in the condenser load. This method is widely used in the United States.
2. The bottom of the condenser is mounted on springs, while the top is connected rigidly to the turbine exhaust nozzle. The springs can be

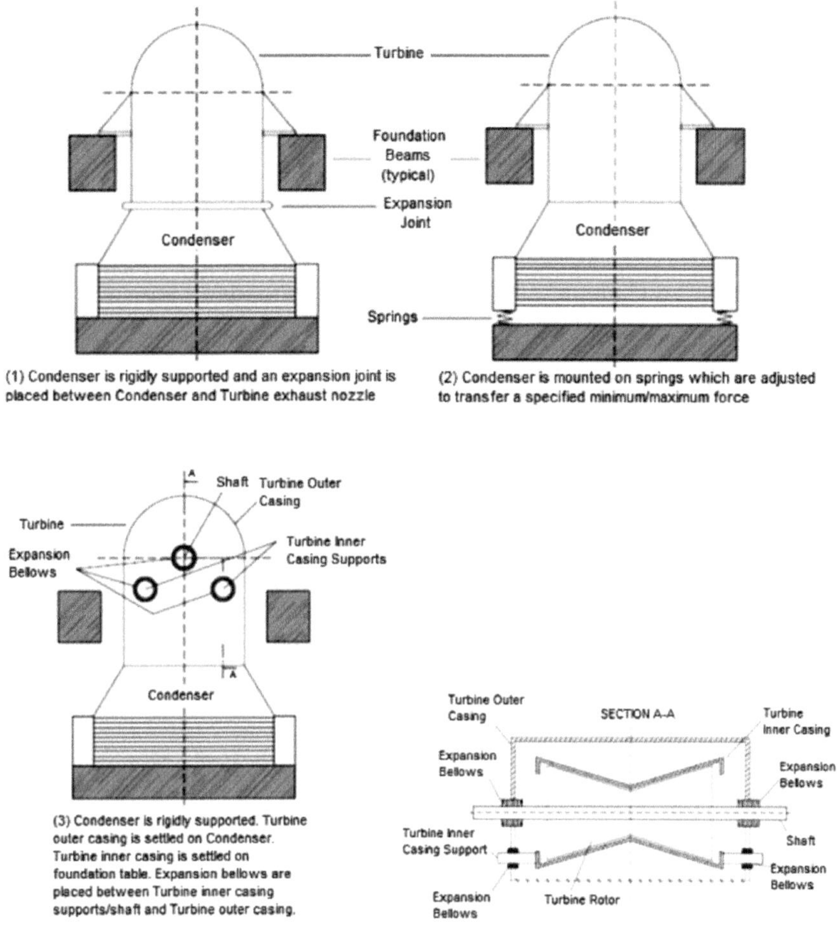

Fig. 2-8. Methods of supporting a condenser
Source: Courtesy of Siemens AG; reproduced with permission.

adjusted to transfer specified maximum and minimum loads to the turbine exhaust nozzle. They can also be adjusted to compensate for load eccentricity, such as from the circulating water pressure load. This method is used worldwide except in the United States.

3. The bottom of the condenser is mounted on rigid supports, while the top is connected rigidly to the turbine exhaust nozzle. To relieve thermal forces and variations in condenser loads, expansion-bellows are used at locations where turbine components intersect with the turbine exhaust nozzle. The condenser loading is predominantly transferred to the foundation basemat at the bottom of the condenser.

2.4 FOUNDATION DESIGN INFORMATION

The design and configuration of foundations for STGs and CTGs are based on subsurface properties of the construction site, as well as the data and criteria provided on the TG and condenser drawings supplied by the manufacturer to the designer at the time of foundation layout design. The manufacturer's drawings provide important information on machine outline and arrangement, foundation loadings, limiting foundation dimensions, piping and penetrations, allowable movements, equipment details, laydown requirements and soleplate, anchor bolt, and embedded plate locations.

2.5 INSTALLATION

The main components of TGs are mounted on soleplates, which are grouted and anchored by bolts to the foundation. The components are designed to slide on these fixed soleplates to minimize thermal stress from expansion and contraction. Guides and/or keys are installed at the component/soleplate interfaces to prevent misalignment of the components from thermal movement.

Each rotor should be aligned and balanced to ensure low vibration during operation. Misalignment of the rotors may result in serious vibration problems and damage to the unit.

2.6 OPERATION

Operating speeds of STGs and CTGs are typically 3,000 or 3,600 revolutions per minute (rpm) for 50 Hz and 60 Hz electrical grids, respectively, except for nuclear STGs. Nuclear STGs typically operate at half speed (1,500 or 1,800 rpm), primarily to reduce the erosions of rotating turbine blades from water droplets, which are more prevalent in a typical nuclear steam cycle. During startup, the machine speed is increased slowly and held at lower speeds to pre-warm the stator shell and to maintain sufficient clearances between the rotor and the stationary components.

Various instruments are provided to continuously monitor unit operation. These instruments can include rotor vibration sensors, differential deflection measurement sensors, and thermocouples. Alarm and trip limits are established for these instruments. Should any of the measurements exceed a limit, the unit will alarm to warn operators or trip the unit to prevent damage to the machine or danger to personnel. The limits may vary significantly for different units, configurations, and manufacturers.

CHAPTER 3

PRELIMINARY TURBINE GENERATOR FOUNDATION LAYOUT AND SIZING

3.1 INTRODUCTION

This chapter presents general descriptions and criteria for the layout and preliminary sizing of STG and CTG concrete foundations, including foundation types, and general design information required, as well as initial sizing guidelines for the tabletop, columns, walls, mezzanine levels, and basemat.

3.2 FOUNDATION TYPES AND GENERAL DESIGN INFORMATION

3.2.1 Foundation Types

Many STG and CTG foundations are heavy concrete structures, whose layouts are primarily determined by the machine manufacturer's foundation interface requirements.

For STG foundations, depending on how the condenser is positioned with respect to the LP turbine, the layout can be categorized as one of two foundation types, for the purpose of this chapter:

- *Elevated Space-Frame Pedestal Foundation:* This arrangement is used with the bottom-exhaust STG layout, for which the foundation has to be elevated to accommodate the height requirement of the condenser. Columns and/or walls are required to support the tabletop where the STG equipment is mounted. The columns and/or walls have a common basemat. Vertical structural elements are sized to meet the stiffness requirements of the tabletop. In many cases, the machine centerline can be as high as 60 ft above the basemat (Fig. 3-1).

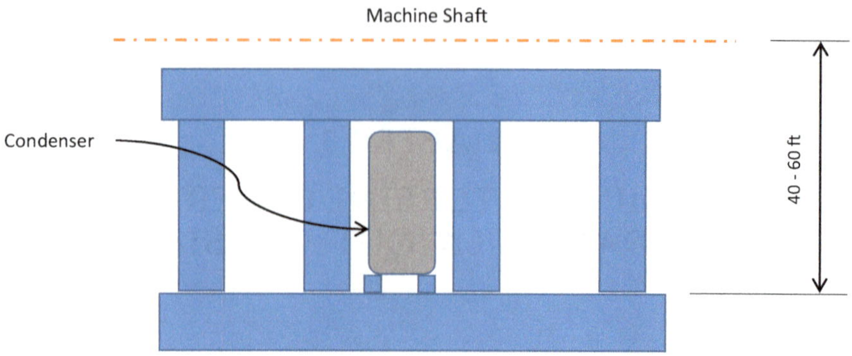

Fig. 3-1. Elevated space-frame pedestal foundation

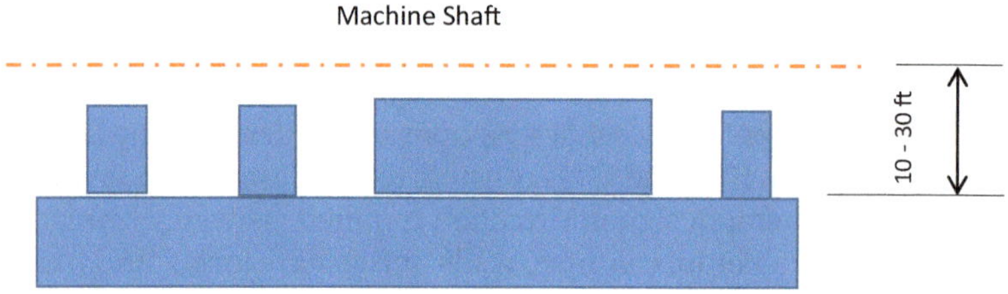

Fig. 3-2. Block foundation

- *Block Foundation:* This arrangement is used with either a side-exhaust or axial-exhaust STG layout. The condenser is either on the side of the LP turbine, or aligned with the TG axial shaft. Individual concrete pedestals, piers, and blocks with a common basemat support the STG equipment. The machine centerline is usually 10 to 20 ft above top of basemat (Fig. 3-2). It should be noted that many low-profile STG-CTG foundations may also have tabletop-type layouts (Figs. 2-3 to 2-5) but can be treated as block foundations when sizing foundation components.

Foundations supporting CTG equipment are also typically block foundations as described previously. In many layouts, the steel columns that support the inlet duct are also supported on the CTG foundation basemat, which results in a larger basemat footprint area. The decision regarding an integrated or isolated foundation basemat for the CTG and inlet duct support columns shall be made through close coordination between civil, plant design, and mechanical disciplines. One important factor to consider in making such a decision is the acceptable vibration levels for the duct support columns.

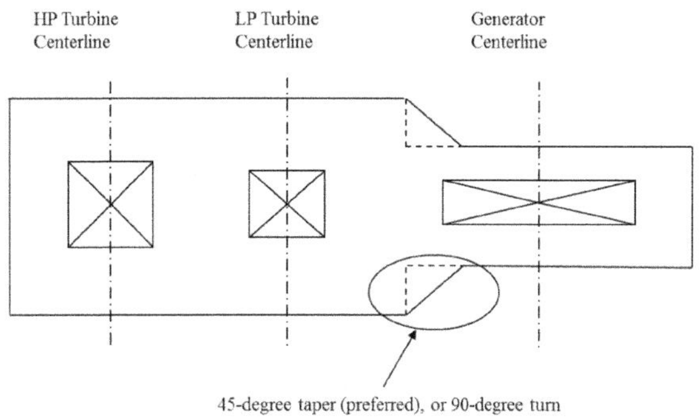

Fig. 3-3. Tabletop layout with a narrower width at generator side

3.2.2 General Foundation Design Information

When initiating the foundation design process, the foundation design engineer should coordinate with other disciplines (e.g., plant design, mechanical, electrical) to collect, review, and understand the following information:

Project Civil/Structural Design Criteria. The Civil/Structural Design Criteria for the project lists the building codes and other design codes required under project contractual agreements.

Seismic design parameters, such as seismic design category (SDC) and spectral accelerations (S_{DS} and S_{D1}) as defined in ASCE 7, should be retrieved from the design criteria. Site-specific seismic studies, if performed, shall be used as the seismic design inputs for the foundation design.

If the foundation is not housed by a turbine building, wind load and snow load design parameters in the design criteria should also be collected.

Machine Manufacturer's Drawings and Documents. The following information, as a minimum, must be provided by the machine manufacturer for the foundation design:

- Machine-foundation interface drawings showing overall foundation and tabletop dimensions, openings, pipe penetrations, grout pockets, anchor bolts, jacking posts, and special embedments.
- Turbine, generator, exciter/collector, and gearbox mechanical outline drawings, which include turbine and generator rotor weights, critical speeds, mass moment of inertia, overall assembly weights, and center of gravity locations.
- Loading documents or drawings showing all applicable design loads from the machine to the foundation, including the dynamic unbalance loads at each bearing, which is the essential design information required to check the foundation vibration level during normal operations.

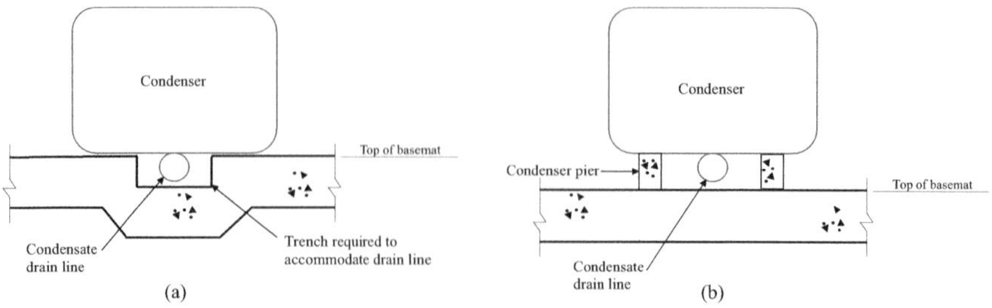

Fig. 3-4. Layout of condenser piers to avoid trench in basemat: (a) Trench in basemat to accommodate drain lines; (b) Concrete piers supporting condenser, eliminating trench in basemat

- For STG foundations, the condenser general arrangement and foundation interface drawings showing condenser layout, foundation interface (anchor bolts and embeds), weights (dry, water test, operational), and applicable loads to the foundation.
- Vendor-specific foundation design and serviceability criteria specifying the unique requirements for the foundation design, such as acceptable vibration amplitudes during normal operation, static and dynamic stiffness requirements, local deflection or misalignment requirements, frequency exclusion requirements, and overall foundation settlement limits.

When reviewing these drawings and documents, the design engineer should perform independent checks to verify that the information provided is complete and reasonable. Following is a recommended list of actions the engineer can take for this purpose:

- Sum up the component weights in the loading table and compare with the total weights presented on the drawings.
- Add up seismic loads at each location to see if the sum matches the total seismic load.
- Use rotor weights to estimate machine normal unbalance forces for both turbine and generator, with appropriate balance grade per ISO1940-1, and compare estimated values with those given in the loading table.
- Review notes on vendor drawings to ensure scopes of supply for anchor bolts and other embed items agree with project contractual agreements.
- Review foundation dimension callouts to ensure there are no missing or conflicting dimensions.
- Review condenser layout drawing to ensure sufficient space is available for installation.

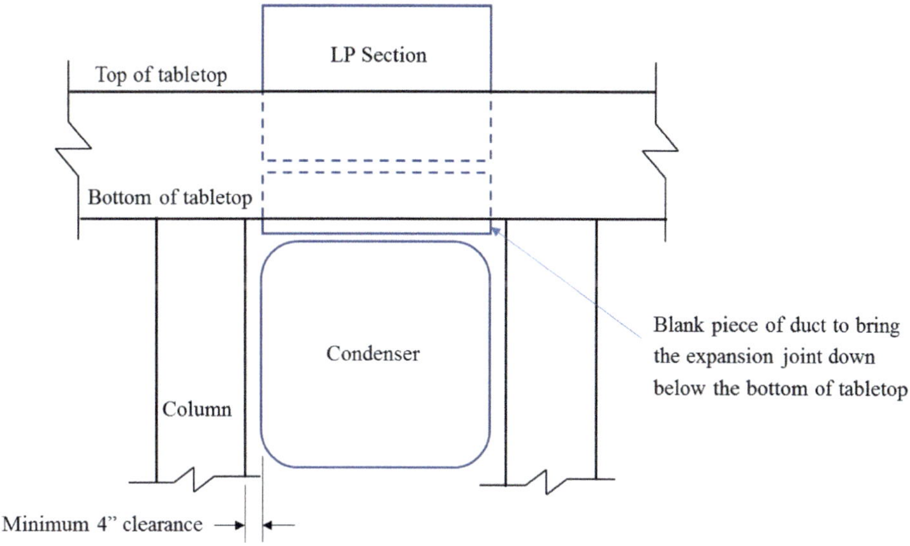

Fig. 3-5. Condenser to LP turbine interface, and minimum space between condenser and columns

Geotechnical Report (Soil or Pile Information). The design engineer should obtain the following site-specific information from the project geotechnical report:

- If supported on soil, the soil allowable bearing capacity, subgrade reaction, density, dynamic and static (i.e., low strain and high strain) Young's modulus, and Poisson's ratio for each soil layer underneath the foundation, up to a depth equal to the longer dimension of the proposed foundation basemat.
- If supported by piles, the pile type, size, length, carrying capacity (with pile spacing reduction factors), and dynamic and static (i.e., low strain and high strain) stiffness values for the piles.

3.3 PRELIMINARY DESIGN: INITIAL FOUNDATION SIZING

When the majority of the information identified in Section 3.2.2 is available and has been reviewed, the design engineer can start working on the preliminary sizing of the foundation. For elevated space-frame pedestal foundations, these include initial sizing of the tabletop geometry and thickness within the manufacturer's boundary limitations, column height, basemat footprint and thickness, and condenser pier sizes. For block foundations, these include the pier and block sizes as well as basemat footprint and thickness.

Though every foundation is unique, previous design experiences with similar foundations supporting comparable equipment are helpful during this preliminary foundation sizing phase.

Because of the massive nature of STG and CTG foundations, it is rare that the overall foundation drifts and local deflections become a design concern. Nevertheless, it is desirable to isolate the foundation, at all levels, from surrounding structures and other foundations with a minimum 1 in. expansion joint to avoid possible vibration transmissions. For high seismic zones, the overall drifts and local deflections should be looked at when sizing the required gap. The gap can be filled with a compressible material at the floor levels.

3.3.1 Elevated Space-Frame Pedestal Foundations

3.3.1.1 Tabletop Geometry. The tabletop geometry is primarily determined by the machine manufacturer's mechanical outline drawings and foundation interface drawings. These drawings define the interface between the machine and the foundation, such as openings, piers, trenches, pockets, anchor bolts, jacking posts, embedded plates, and special embedments. Except for machine layouts where a gearbox is present to change a higher turbine speed to a lower generator speed, the tabletop should be symmetric about the machine longitudinal shaft axis. The design engineer should thoroughly review these drawings to ensure all the aforementioned information is present on the drawings.

The manufacturer's foundation drawing often shows a suggested tabletop thickness. However, if not given, a trial thickness of 6 ft (for power output less than 100 MW) and up to 11 ft (for power output of more than 800 MW) can be used. Unless specifically required by the manufacturer, the bottom of the tabletop should be maintained at the same elevation to simplify the geometry and facilitate construction.

Some STG manufacturers require individual piers on top of the tabletop to support the equipment. These piers are usually sized by the manufacturer and are shown on the manufacturer's drawings.

Segregated by openings and supporting columns and/or walls, most tabletop structural members are considered "deep beams" per ACI 318 definitions. The members running parallel to the machine shaft are called "longitudinal girders," whereas the members running transverse to the machine shaft are called "transverse beams." The clear span-to-depth ratios for all these members are in the range of 1.5 to 3.5 but should not be larger than 4.0. The member depth-to-width ratios are in the range of 1.0 to 2.0.

For some machine designs, the generator equipment is much narrower than the turbine equipment. If this is the case, the width of the tabletop in the area of the generator may be reduced for economic considerations. The transition from the wider turbine tabletop to the narrower generator tabletop can either be a 45-degree taper (preferred) or a 90-degree turn (see Fig. 3-3).

The perimeter of the tabletop may have a minimum 4 in. overhang from faces of the columns to simplify reinforcement placement (i.e., to avoid interference between column vertical reinforcement and tabletop side face reinforcement).

3.3.1.2 Columns, Wall, and Condenser Piers. Depending on the size of the STG equipment, there may be six or more columns supporting the tabletop. For a layout with two LP turbines, with one condenser under each, a transverse wall separating the two condensers is often required to provide sufficient stiffness for the tabletop to resist large vacuum loads. The columns and wall should have a common basemat.

The required clear height of the columns, between top of basemat and bottom of tabletop, should be determined by the condenser dimensions and required construction clearances. The foundation design engineer should review the condenser general arrangement (GA) drawings, along with the STG manufacturer foundation drawings, and coordinate closely with other disciplines, STG manufacturer, and condenser manufacturer to determine the required height of the tabletop foundation. The overall layout of the vertical structural elements should be symmetric about the machine shaft axis.

If possible, columns should be located underneath intersections of tabletop girders and beams. However, column locations can be adjusted to allow for sufficient space for the condenser or other equipment and pipe routings. The clear space between adjacent columns and an installed condenser should be at least 4 in. (see Fig. 3-5). For cases where there is a wall separating two condensers, the wall thickness is limited by the minimum required clearance as well, though a thicker wall might be more structurally desirable.

Most columns should have a square or rectangular shape. Some may have an "L" shape if the tabletop layout has a transition from a wider turbine tabletop to a narrower generator tabletop, as described in Section 3.3.1.1. Based on surveys of past large STG pedestal foundations, the sizes of the columns are always governed by stiffness requirements and dynamic performance considerations rather than strength requirements. The actual sizes of columns vary, depending on the weight and size of the equipment they support. Typical columns have a shorter dimension of 5 to 6 ft and can be up to 12 to 14 ft in the longer dimension. When preliminarily sizing column cross sections, it is desirable to have similar compressive stress levels resulting from dead load only at the base of each column to minimize the differential settlement of the foundation.

If the condenser is set at the top of the foundation basemat, a trench in the basemat will be required to accommodate the condensate drain lines that exit the bottom of the condenser [Fig. 3-4(a)]. Such a trench complicates the basemat design by interrupting reinforcement and compromising strength. It also complicates formwork and excavation planning. A preferable alternative is to set the condenser on piers to allow for the condensate drain lines to pass above top of basemat [Fig. 3-4(b)]. However, coordination with construction is necessary before choosing this alternative, as rolling-in of the condenser may not be possible if the piers are too high.

The location of the large diameter circulating cooling water pipes must also be examined. In many cases, the pipes penetrate below grade elevation adjacent to the basemat. Depending on the location of the penetration, the basemat may have to be notched to accommodate the vertical pipe. Such notches complicate reinforcing and formwork. This item should be reviewed with the mechanical and plant design disciplines early in the design phase, so that measures may be taken to avoid notches in the basemat.

The interface between the LP exhaust hood and the condenser neck requires special attention. This interface often has an expansion joint. If the interface falls within the depth of the tabletop transverse and longitudinal beams under the LP turbine, the interior bottom edge of these beams may have to be notched to accommodate the installation, visual inspection, and maintenance activities for the expansion joints. Such notches complicate formwork and the beam reinforcing steel arrangement. To remedy this situation, a blank piece of duct may be added between the LP flange and the expansion joint to lower the elevation of the expansion joint to below the bottom of the tabletop concrete (Fig. 3-5).

3.3.1.3 Mezzanine Platforms and Beams. Some foundation designs require a mezzanine level platform to provide access and for laying out small, miscellaneous equipment, particularly under the HP turbine and generator. To support the mezzanine level, either steel beams with grating or concrete beams with a concrete slab may be used. Concrete beams spanning between the columns may have section depths ranging from 3 to 5 ft.

3.3.1.4 Condenser Piers. As discussed in Section 3.3.1.2, condenser piers may be required to avoid a trench in the basemat. Some condenser manufacturers may also require piers. Condenser piers should be sized per condenser manufacturer's interface requirements. When cast-in-place bolts are used, the anchor bolt locations, along with the condenser installation procedures, should be reviewed to ensure that the bolts will not block the installation of the condenser. As an alternative, anchor bolts with couplers can be used to help reduce interferences during condenser installation.

3.3.1.5 Basemat. Once the overall tabletop dimensions and the column locations and their initial sizes are determined, the footprint of the basemat can be sized. A rectangular shaped basemat is recommended, and the edges of the basemat should extend 1 to 3 ft from the faces of the columns.

The basemat footprint dimensions must be sufficient to provide support for the turbine pedestal columns and satisfy all stability and soil bearing or pile requirements. The horizontal eccentricity, in either direction, between the center of gravity of the machine-foundation system and the centroid of the foundation basemat, should not exceed 5% of the corresponding basemat dimension. This is to reduce the number of possible complex coupled mode shapes of a foundation, as well as to minimize potential foundation differential settlement.

The basemat is typically 5 to 8 ft thick, but can be up to 12 ft thick for heavier STG equipment with an output of more than 800 MW. However, the thickness should not be less than 1/5 of the largest clear span between the columns in the longitudinal direction. The basemat should be symmetric about the machine shaft axis.

Isolation of the basemat from the surrounding turbine building foundation is preferred, and may be required by the machine manufacturer. However, for sites with high seismic loads, a combined foundation may be required for resisting lateral loads and minimizing horizontal displacements. In cases where turbine building columns are located on the STG pedestal basemat, vertical bracing should not be framed into the base of these columns. This helps reduce possible transmission of machine vibration to the turbine building structure. If vibration of the building columns becomes a concern, a vibration isolation system (VIS) may be used.

In addition to the load transfer and frequency control, the basemat must be sufficiently stiff to minimize differential displacements of the foundation. If differential settlement is a potential problem, a vibration isolation system for the tabletop is a possible solution. Information on spring-mounted foundations is provided in Chapter 10 of this book.

Basemat should be embedded in soil, rather than sitting on top of grade or being surface-mounted. Frost depth requirement should be considered when determining the embedment depth. The soil embedment will help reduce the dynamic responses of the foundation because of increased soil damping effects. The effectiveness of the surrounding soil is contingent on suitable engineering compaction practices. The top 6 to 12 in. portion of the backfill soil may be loose and should not be considered as effective confinement. In addition, cohesive backfills may experience separation from the embedded basemat resulting from consolidation and desiccation. In such cases, a reduction of embedment should be considered.

3.3.1.6 Rules of Thumb for Preliminary Elevated Space-Frame Pedestal Foundation Sizing. Once the preliminary geometries of the tabletop, columns, walls, and basemat are determined, the following additional rules of thumb are useful with respect to the overall foundation sizing. These empirical rules and the guidelines provided in this chapter offer a good starting point for the foundation design. However, it is up to the design engineer's engineering judgment to decide whether to meet some or all these rules for a given foundation, and these preliminary sizes are subject to change per detailed analysis and design. Adequate foundation performance can be achieved with designs that deviate from these rules. See Fig. 3-6 for a graphical summary.

1. The tabletop concrete weight should not be less than the total machine weight.
2. $W/A \approx 60$ to 90 psi
   where
   W = Combined machine and tabletop weight
   A = Cross-sectional area of vertical structural elements (columns and walls) that support the tabletop.
3. The self-weight of all vertical structural elements (columns and walls) supporting the tabletop should be 40 to 60% of the combined weight of the machine and tabletop.
4. The basemat weight should not be less than the combined machine and tabletop weight.
5. The total foundation concrete weight (tabletop + columns/walls + basemat) should not be less than 3.5 times the machine weight.
6. The maximum "h/r" ratio of the columns should not be larger than 22, the limiting value given in ACI 318 when the P-delta effect has to be accounted for. "h" is the clear height between top of basemat and bottom of tabletop and "r" is the least gyration radius of the columns. For rectangular columns, the gyration radius is equal to 0.29 times the least dimension of the column cross section.
7. The basemat thickness should not be less than (L/30) plus 2 ft, where "L" is the length (in ft) of the basemat, or not less than 1/5 of the largest clear span between the columns in the longitudinal direction.
8. The basemat width (B) should not be less than the vertical distance from the bottom of basemat to machine shaft centerline.

9. The average soil bearing pressure, or pile compression, should not exceed 3/4 of that allowable under self-weight plus normal dynamic loads.
10. If the foundation is pile-supported, the pile distance between the outer rows in the transverse direction should not be less than the height measured from bottom of basemat to machine centerline.

3.3.2 Block Foundations

3.3.2.1 General Discussions. Concrete block foundations are typical for axial-exhaust and side-exhaust STGs and CTGs. They are composed of a common thick basemat with individual piers and blocks supporting the machine components.

Axial-Exhaust STG Foundation: The foundation for an axial-exhaust STG layout requires a low-profile block foundation rather than an elevated space-frame pedestal foundation. Consideration needs to be given to the turbine condenser connection. The first support pier for the air-cooled condenser may be subject to a large vacuum load, which makes an isolated pier difficult to design. Therefore, it is common practice to integrate this first support pier with the STG block foundation. If an isolated pier foundation is used, close coordination with the condenser supplier is required to ensure the deflection owing to the large vacuum load during normal operations is appropriately addressed.

Side-Exhaust STG Foundation: Similar to the axial-exhaust STG foundation, a low-profile block foundation is required for the side-exhaust STG. The side-exhaust STG foundation is slightly taller than the axial-exhaust configuration. The mat for the condenser foundation can be integrated with the STG foundation basemat to minimize possible differential settlement issues. Based on the turbine condenser connection, a large vacuum load may also be a major foundation design consideration, similar to the axial-exhaust configuration.

CTG Block Foundation: A CTG foundation is a block foundation consisting of a massive concrete basemat and low-profile piers and blocks. In some designs, the footprint of the basemat is enlarged to include the steel inlet duct support columns and the first supports of the exhaust duct.

3.3.2.2 Preliminary Block Foundation Design Guidelines. In general, the following guidelines can be followed for preliminary sizing of STG and CTG block foundations:

1. Machine manufacturer's recommendations for the foundation should be followed.

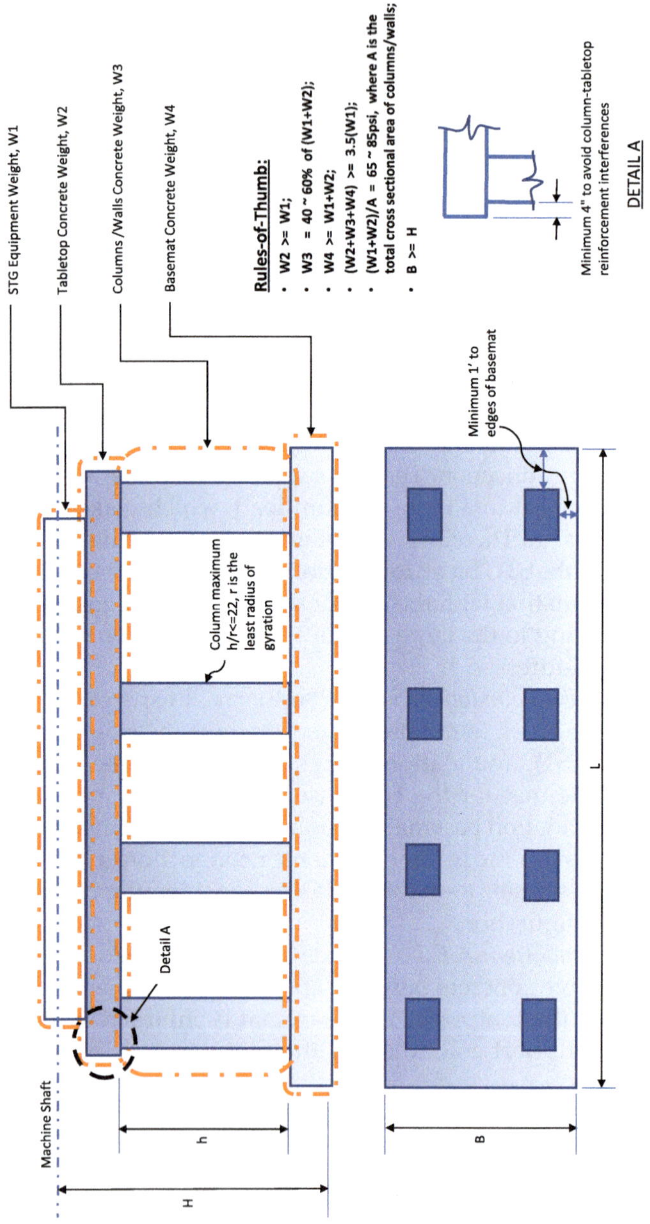

Fig. 3-6. Rules of thumb for elevated space-frame STG pedestal foundations

2. The foundation width should be at least 1.2 times the vertical distance from bottom of basemat to the machine centerline.
3. The total concrete self-weight should be at least 2.5 times the total equipment weight.
4. The basemat thickness should not be less than 3.5 ft, preferably 4 to 6 ft, depending on the weight of the equipment. Similar to basemat for an elevated space-frame pedestal foundation, it should be embedded in soil (see Section 3.3.1.5).
5. To be considered a rigid block, the basemat thickness should not be less than 1/5 of the least foundation dimension or 1/10 of the largest foundation dimension. Foundation basemats not meeting this criterion should be treated as flexible in the dynamic analysis. Note that the rigidity of a basemat also depends on soil stiffness and machine excitation frequencies. A detailed finite element analysis may be required to confirm such a "rigidity" assumption.
6. The center of mass of the foundation-machine system should coincide with the centroid of the soil or pile resistance. The horizontal eccentricity should be limited to 5% of the corresponding foundation dimension.
7. To avoid local resonance, the fundamental natural frequencies of individual piers, calculated assuming fixed base condition, should be checked to ensure they are higher than the machine operating frequency by at least 40%. The uncracked section properties should be used in such frequency analysis, since the stress level is expected to be very low.

3.4 PILE LAYOUT

Because of the heavy downward loads on large TG foundations, a high allowable soil bearing capacity, usually 4,000 psf or higher, is required. For site subsurface conditions that do not meet this requirement, or for other geotechnical considerations (e.g., uniform settlement control or possible soil liquefaction in high seismic regions), the geotechnical report may recommend a pile-supported deep foundation.

There are various types of piles that are used in practice, such as prefabricated concrete piles, cast-in-place or auger-cast concrete piles/caissons, and steel H-shape or pipe piles. With respect to pile working mechanisms, piles can be categorized as

- Friction piles that rely on friction capacity between pile skin and soils;
- Bearing piles that rely on direct end bearing capacity on bedrock or the stiff soil below the pile tip; and
- Piles that rely on both friction and direct end bearing mechanisms.

The connection between the piles and the basemat is categorized as either "fixed-head" or "free-head." Whereas the pile connection has little effect on the pile compression and tension capacity, "fixed-head" piles will have a higher lateral capacity than "free-head" piles. For cases where lateral loading demands (e.g., owing to seismic loads) are high, a "fixed-head" pile may be required.

The pile lateral capacity reduces for center-to-center pile spacings of less than 6D, where D is the width or diameter of the pile. The geotechnical engineer should provide different reduced pile lateral capacities for spacings ranging from 3D to 6D. For a specified lateral movement, piles in the leading row of a pile group with 3D spacing can experience a reduction of 10 to 15% in allowable lateral load, whereas the corresponding reduction for piles in trailing rows can be 35 to 45%. Usually, no reduction in lateral capacity for pile spacings over 6D is required. Conversely, the pile axial capacity is usually not affected by pile spacing. For purely friction piles, however, the potential danger is that the piles may fail as a group rather than individually if they are too closely spaced (less than 3D). The foundation design engineer should consult the geotechnical engineer for using appropriate pile capacity reduction factors.

For TG foundations, the allowable lateral pile load usually governs the number of piles required and the pile layout. When determining the pile layout, a uniform layout pattern, either rectangular or staggered, is preferred, with a minimum 4D center-to-center pile spacing.

A calculation assuming a rigid basemat/pilecap may be performed to estimate the average demands (compression, tension, and shear) on the piles. The loads should include dead loads, machine emergency or accidental loads, and seismic loads. For elevated space-frame pedestal foundations, the total seismic lateral loading demand on piles can be calculated as the square root of the sum of the squares (SRSS) value of the superstructure seismic base shear and the basemat seismic shear (ASCE 2011, Section 5.3.5).

REFERENCES

ACI (American Concrete Institute). (2014). "Building code requirements for structural concrete and commentary." *ACI 318-14*, Farmington Hills, MI.

ASCE. (2010). "Minimum design loads for buildings and other structures." *ASCE/SEI 7-10*, Reston, VA.

ASCE Task Committee on Seismic Evaluation and Design of Petrochemical Facilities. (2011). *Guideline for seismic evaluation and design of petrochemical facilities*, 2nd Ed., ASCE, Reston, VA.

ISO (International Organization for Standardization). (2003). "Mechanical vibration—balance quality requirements for rotors in a constant (rigid) state, Part 1: Specification and verification of balance tolerances," 2nd Ed., *ISO 1940-1*, Geneva, Switzerland.

CHAPTER 4

FOUNDATION LOADS AND LOAD COMBINATIONS

4.1 INTRODUCTION

TG foundations must be designed to have adequate structural integrity to withstand all the loads imposed during the service life of the equipment, from both strength and serviceability (deflection) standpoints. Such loads are produced not only by the operating turbine and generator, but also from pipes, valves, enclosures, other associated equipment, and environmental forces such as earthquake and wind. Actual loads on a TG foundation vary from machine to machine. This chapter aims at defining the types of loads generally considered in the design of a TG foundation and not specific to any manufacturer's machine.

The TG foundation loads can be divided into three categories:

1. Foundation Self-weight and Equipment Loads—Equipment loads produced by the TG, including machine weights, normal operation loads, emergency operation loads, and catastrophic equipment loads (such as loss of turbine blade).
2. Environmental Loads—These loads include seismic, wind, and snow loads.
3. Installation and Maintenance Loads—These loads include equipment laydown and rotor removal loads, as well as live loads on the tabletop and basemat.

Building codes (e.g., ASCE 7) stipulate the load combinations for structural design using categories such as dead load and live load, for which different load factors shall be used. TG foundation design loads falling within categories 1 and 3 as listed can be further categorized into either dead or live loads as appropriate for structural design purposes. See

Section 4.9 for details on strength design load combination considerations. Serviceability load combinations are covered in Chapter 7.

4.2 FOUNDATION AND EQUIPMENT SELF-WEIGHTS (DEAD LOADS)

The foundation dead load includes the self-weight of all supporting structural members of the foundation.

The machine dead load is the static load at each TG support based on the final installed weight of the TG. Machine manufacturers provide this load as reactions on the foundation in a loading table.

4.3 EQUIPMENT LOADS DURING NORMAL OPERATION

Normal operation loads are produced by the TG during normal operating conditions. Continuous operation of the TG at normal operating load levels should have no adverse effect on the machine life and performance. The foundation system needs to be designed so that any degeneration of the foundation resulting from operational loads does not have long-term detrimental effects on the equipment. These loads are provided by the machine manufacturer.

Sections 4.3.1 to 4.3.8 describe general operational loads produced by the TG equipment.

4.3.1 Normal Operational Torque Load

Fluid or magnetic coupling between the rotor and stator can produce a torque oriented about the turbine generator rotor axis (Fig. 4-1). The normal operational torque load is based on the highest expected output of the turbine generator or the maximum generator rating. For both steam and gas turbines, the normal operational torque load is oriented in the opposite direction from the unit rotation. For the generator, the normal operational torque load is oriented in the same direction as the unit rotation. When looking in the direction of gas flow, the gas turbine rotates counterclockwise. For single-shaft applications, the direction of rotation for the steam turbine must be the same as that of the combustion turbine.

4.3.2 Normal Machine Unbalance Load

In practice, it is impossible to balance the TG rotor to eliminate all unbalance forces during rotation; therefore, some unbalance will always exist. The magnitude of the machine unbalance forces depends on factors determined by

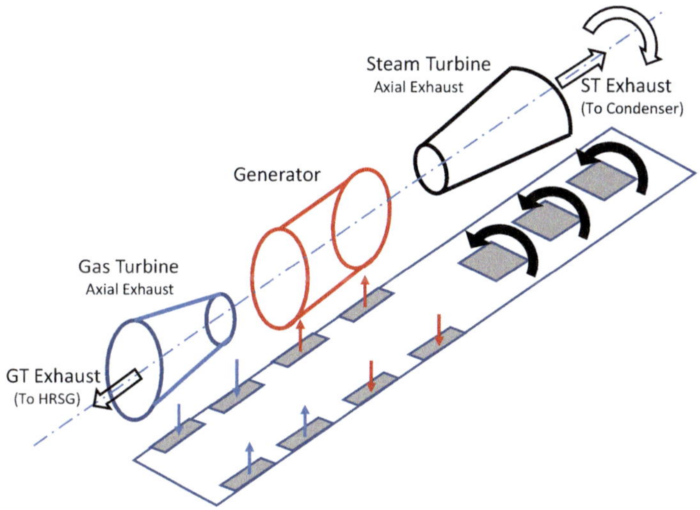

Fig. 4-1. Normal operational torque direction for a single shaft STG/CTG arrangement
Source: Courtesy of GE Power; reproduced with permission.

design, manufacturing, installation, and maintenance procedures. These factors may include an axis of rotation that does not pass through the center of mass of the rotor, deflection of the shaft owing to gravity, uneven thermal expansion, misalignment during installation or corrosion, and wear of moving parts. The cumulative effect of these factors results in the presence of unbalance forces that are synchronous with the shaft rotational speed. These forces are transmitted to the foundation through the shaft bearings.

Dynamic Forcing Function. For the purpose of defining the normal machine unbalance dynamic load, the TG rotor is approximated as a multi-supported shaft with a rotating disk corresponding to each of the turbine stages and the generator as shown in Figs. 4-2 and 4-3. The forcing function $F(t)$ corresponding to each of the turbine stages and the generator is

$$F_i(t) = M_i \frac{G\Omega^2}{\omega} \sin(\Omega t + \alpha_i) \tag{4-1}$$

where

M_i – the mass of the rotating mass i;
$G = e\omega$, balance grade, a measure of the balance quality grade of the rotor;

e = the rotating mass eccentricity, which equals the distance between the center of rotation and the mass center of gravity of the rotating mass (Fig. 4-3);

ω = the machine design operating speed in radians/second;

Ω = the rotational speed in radians/second for which the unbalance force is being computed;

α_i = the relative phase angle of the unbalance for rotating mass i;

t = time.

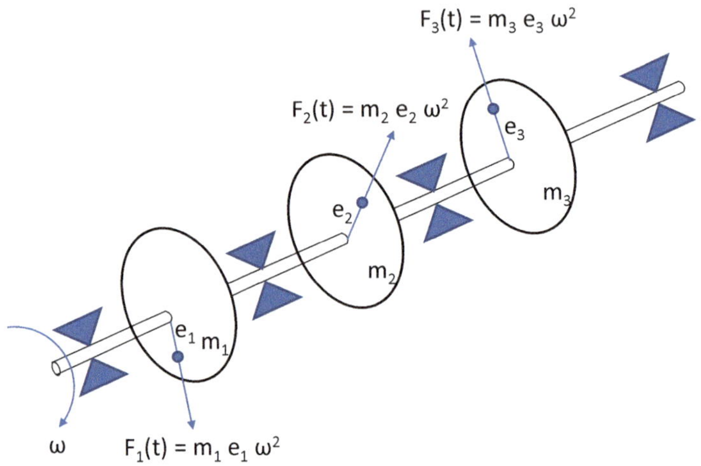

Fig. 4-2. Unbalance forces along shaft with multiple supports

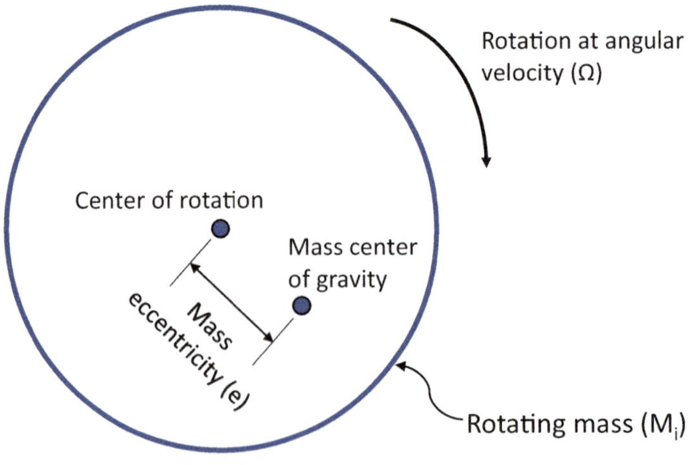

Fig. 4-3. Rotating eccentric mass

For TG sets, the balance quality grade per ISO 1940-1 is G2.5. This corresponds to a G value (or $e\omega$ product) of 0.1 in./s (2.5 mm/s). The relative phase angle α_i of the unbalance forcing function, corresponding to each stage of the turbine and generator, is random and is unknown at the design stage. However, its random nature should be considered in the determination of the probable maximum response of the TG foundation system. The TG manufacturer may perform a specific rotor dynamic analysis of the power train and provide the unbalance loads that can be used in lieu of Eq. (4-1). Section 7.2 discusses how to combine unbalance loads at different bearings to produce the probable maximum responses for design purposes.

The 0.1 in./s (2.5 mm/s) G value corresponds to a minimum operating unbalance, and the actual operating unbalance varies and is likely to be larger over time depending on machine wear and maintenance schedule. The design unbalance should be specified by the TG manufacturer. A G value of 0.25 in./s (6.3 mm/s) may be used when the manufacturer's information is not provided.

4.3.3 Axial Emergency Thrust or Pressure Differential Load

This load has two components: the axial load from flow momentum change through the gas turbine, and the axial load from the pressure differential between the gas turbine inlet and exhaust, as shown in Fig. 4-4. This foundation load may be provided by the manufacturer as a static operational load corresponding to turbine operation with a normal level of exhaust back pressure, or as a static emergency load corresponding to turbine operation with an emergency shutdown level of exhaust back pressure.

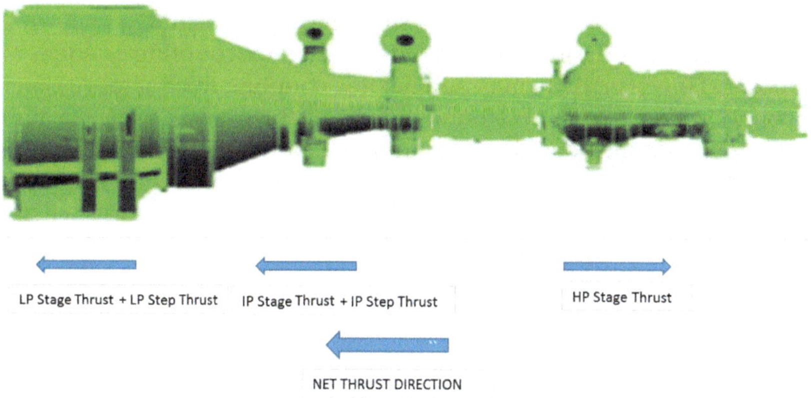

Fig. 4-4. Axial emergency thrust or pressure differential load
Source: Courtesy of GE Power; reproduced with permission.

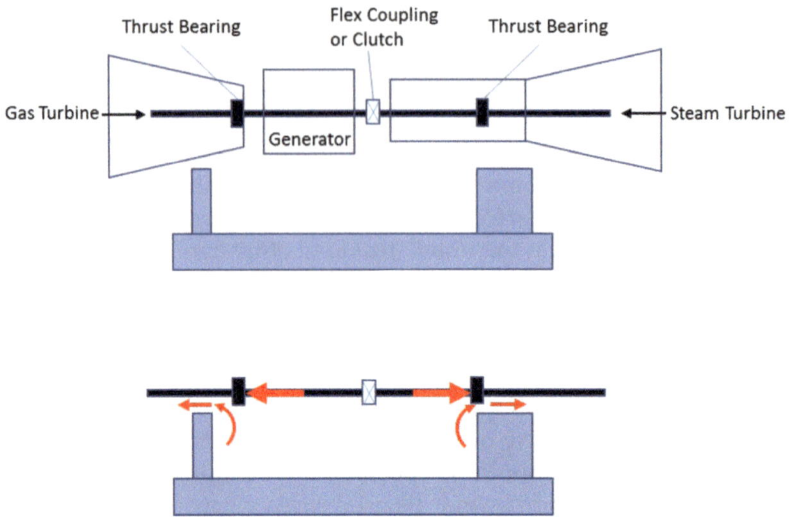

Fig. 4-5. Flex coupling or clutch force (flex coupling or clutch in compression—operating state)
Source: Courtesy of GE Power; reproduced with permission.

4.3.4 Axial Thrust Flex Coupling

This is an axial load that may exist in rotor trains that include a flex-coupling or clutch, as shown in Fig. 4-5. The flex-coupling load may differ in magnitude and direction between the cold-set and operating state due to installation pretensioning or compression and subsequent rotor movement to the hot operating state. The flex coupling load will not impart a net resultant load to the foundation, since the load transmitted to the thrust bearing and foundation on each side of the flex coupling is equal and opposite. However, there may be a net flex-coupling force across a specific foundation element. The TG foundation load tables provided by the manufacturer should include this flex-coupling load when applicable. Depending on the location of the thrust bearing and type of unit, the flex-coupling load may either counteract or be additive to other axial forces such as jet thrust, condenser vacuum, shutdown level backpressure, and axial seismic accelerations affecting the rotor mass.

4.3.5 Thermal Loads

Temperature changes in the turbine and in the generator cause expansions and contractions of machine parts to occur, forcing various parts to slide. As the machine heats up, the entire shaft expands. This expansion does not impart any load on the foundation, since the entire shaft system is fixed longitudinally by a single thrust bearing and the shaft slides freely across the well lubricated journal bearings.

However, unlike the expansion of the shaft, the heat buildup in the casings can impose thermal loading on the foundation. The HP and IP turbine casings are supported on soleplates located on the top of the STG tabletop transverse beams. The LP turbine casings, the generator casing, and the exciter are supported on soleplates located on the longitudinal and transverse beams of the STG tabletop. An example of where the casings are anchored to the foundation in the longitudinal and transverse directions and their corresponding thermal loading is shown in Figs. 4-6 and 4-7.

The TG foundation may also be subjected to thermal loads resulting from ambient environmental temperature variations. The design engineer should work with the manufacturer when determining what type of thermal effects should be considered in a foundation design.

Thermal Friction. Thermal friction loads can be developed at the turbine generator supports from static frictional resistance to thermal movement of the turbine generator. Thermal friction loads normally originate at the weight-bearing turbine generator supports and impart corresponding reaction loads on the supports that maintain the horizontal alignment of the turbine generator, for example, anchor supports and centerline guides. The turbine generator thermal friction loads will not impart a net resultant load on the foundation, since the load at one support is balanced by an

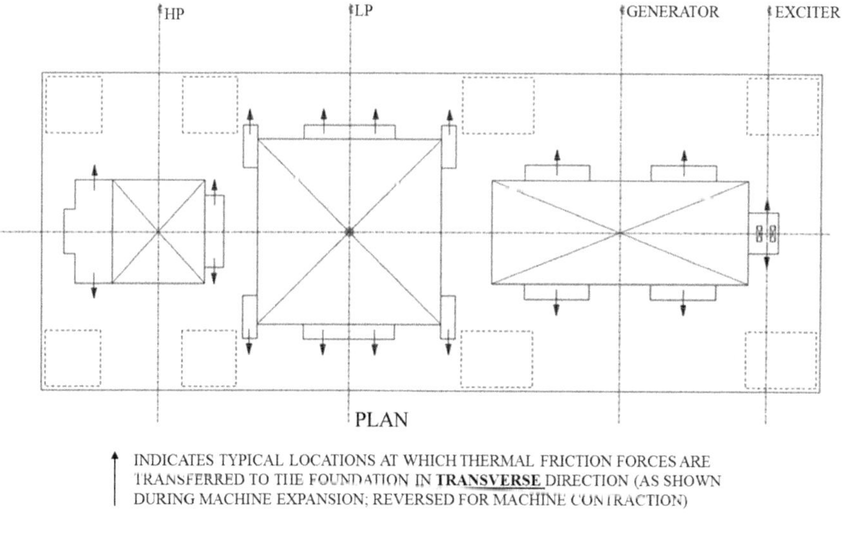

Fig. 4-6. Thermal friction loads at tabletop (transverse)

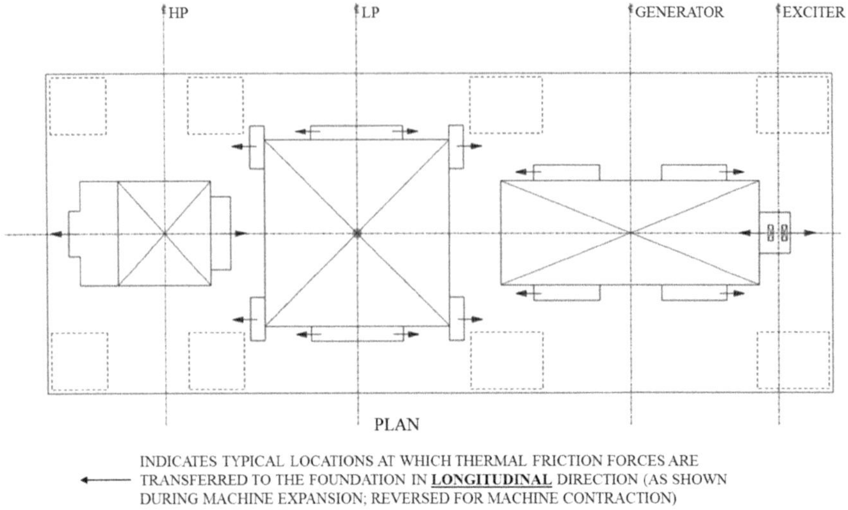

Fig. 4-7. Thermal friction loads at tabletop (longitudinal)

equal and opposite load at another support. The tabulated thermal friction loads provided by the manufacturer are the design loads for each specific support location, and are not intended to be collectively applied to the foundation as tabulated. The thermal friction loads for each load case must be applied not to impart a net resultant load on the foundation in order to be a valid load case. There can be a thermal friction force across a specific foundation element as depicted in Figs. 4-6 and 4-7, but any system of thermal friction loads applied to the equipment interface must be in equilibrium and not produce a resultant shear or overturning moment acting on the foundation.

Thermal Expansion. Thermal expansion loads are the loads that can be developed at the turbine generator supports from mechanical resistance to the thermal movement of the turbine generator. Examples include steam turbine flex leg supports, which deflect to accommodate thermal movement of the turbine. Thermal expansion loads are similar to thermal friction loads, with the fundamental difference that a thermal friction load ceases to exist once the turbine generator moves (i.e., the static resistance force is overcome), whereas a thermal expansion load is always present once the turbine generator moves from the unrestrained position. Similar to thermal friction loads, the thermal expansion load taken collectively at all the turbine generator supports will not impart a net resultant load on the foundation, though a thermal expansion force can exist across a specific foundation element as described for thermal friction loads.

During the startup condition, the direction of the transverse expansion loads acting on the tabletop will be outward from the centerline of the turbine, while during the shutdown condition the loads will act inward to the centerline of the turbine. In the case of the longitudinal expansion loads, an assumption may be made as to the direction of forces on the soleplate, and any unbalance forces between the two adjacent anchorages that prevent longitudinal movement of the turbine may be applied as a concentrated load at the anchorage points.

The total thermal loading is a vectorial sum of the longitudinal expansion loads parallel to the shaft axis and the transverse expansion loads perpendicular to the shaft axis (i.e., vectorial sum of the loads indicated in Figs. 4-6 and 4-7).

The magnitude of the thermal loading under any soleplate is provided by the machine manufacturer.

Thermal Gradient in Foundation from Operation. The stresses and deflections caused by the thermal gradient within the foundation caused by the elevated temperatures of surfaces near the equipment are generally not considered in the design of the foundation, unless explicitly required by the manufacturers. The additional deflections from thermal expansion of the foundation may be significant; however, quantitative evaluations of these deflections are not typically made, in part because the results of calculated stresses and deflections caused by thermal gradients yield inconsistent results. These results are primarily influenced by the following factors:

- Cement type (I, II, III, IV);
- Coarse aggregate type;
- Boundary condition restraint;
- Heat dissipation through mass concrete;
- Linear coefficient of expansion of the foundation materials;
- Thermal conductivity of the foundation materials;
- Specific heat of the foundation materials;
- Concrete density; and
- Element geometry of the foundation model.

The expected thermal deflection at various bearings is estimated by the manufacturer, based on past field measurements on existing units. The machine erector can then compensate for the expected deflections during initial alignment. Successful past performance of turbine generator foundations provides ample justification for such an approach.

Environmental Thermal Effects. The stresses and deflections caused by thermal loading of the foundation from environmental effects should be considered in the design. Applicable situations include, but are not limited to, the following:

- Outdoor units in which there is a large temperature change between day and night or between seasons.
- Outdoor units in which a temperature differential exists as a result of the sun shining on only one side or part of the foundation.
- Outdoor units with CT protected by an enclosure where the interior concrete foundation is in contact with high and uneven air temperatures caused by inadequate design or malfunctioning heating ventilation system.
- Indoor units in which there are temperature gradients on the foundation resulting from cold or hot air blowing over parts of the foundation.

Some turbines develop greater than normal thermal conditions, with concrete surface temperatures exceeding 150°F (66°C) around and within the foundation. This is especially true for outdoor units when a combustion turbine is located and protected by an acoustical enclosure. The air space in the enclosure can be overheated when ventilation is not designed or functioning properly. The engineer should address the effects of these thermal conditions in the design phase per the turbine manufacturer's requirements. Inadequate consideration of the thermal effects can lead to early cracking of the foundation, which can be intensified by dynamic effects. Calculation of thermally induced bending requires proper determination of the heat distribution through the thickness of the foundation. ACI 307 provides certain guidance that can be extrapolated to hot equipment. ACI 349.1R also provides methodologies that may be transferable to certain machine foundations. When a high temperature (250–400°F) is specified for operating conditions by the turbine manufacturer, cracked section properties may need to be used to reflect the expected cracking under such a temperature, as required.

Methods for mitigation of the thermal effects include

- Providing sufficient insulation between the hot equipment and the concrete;
- Providing sufficient cover to the reinforcement so that thermally induced cracking neither degrades the bond of the reinforcement nor increases the exposure of the reinforcement to corrosives;
- Providing sufficient reinforcement to control the growth of thermal induced cracks;
- Providing high-temperature resistant concrete by selecting suitable materials such as aggregate, cement, and admixture; epoxy-coated rebar and epoxy-based grout should not be used for high-temperature

conditions (over 120°F), as the epoxy may not function properly in high temperatures; and

- Improving the CT enclosure mechanical system design so that the enclosure and concrete temperature can be lowered to less than 200°F.

4.3.6 Condenser Vacuum Load

When an expansion joint is provided between the condenser and the turbine exhaust nozzle, the difference between the atmospheric pressure on the casing of the turbine and the vacuum pressure inside the condenser results in a force on the turbine. This vacuum load can be as large as several times the weight of the condenser. Typically, one condenser unit is provided for each LP turbine, and its vacuum load is transmitted to the foundation through the turbine soleplates. The vacuum in the condenser applies a force on the turbine, pulling it toward the condenser. The condenser vacuum loads are normally based on zero absolute pressure, with a flexible connection to the condenser. The exact distribution of this loading is provided by the turbine manufacturer. For the case where the condenser is rigidly attached to the turbine exhaust nozzle, no vacuum load is transmitted to the STG foundation. Fig. 2-8 illustrates the common methods of supporting a down exhaust condenser and Fig. 4-8 illustrates a typical condenser vacuum loading scenario for a side exhaust arrangement.

4.3.7 Condenser Operational Dead Load

When the bottom of the condenser is mounted on rigid supports, the dead weight of the condenser is transmitted to the basemat of the STG foundation. When the condenser is welded to the exhaust nozzle and supported on springs, the proportion of the condenser dead load distributed between the foundation tabletop and basemat depends upon the stiffness and initial setting of the springs supporting the condenser. The loading is provided by the turbine and/or condenser manufacturer. Additional information on spring-mounted condensers is provided in Chapter 10.

4.3.8 Piping and Valve Reactions

Pipe forces result from pipe weight, wind acting on the pipes, water hammer caused by rapid valve closure, cold springing, fluid dynamics, seismic forces, and thermal expansion.

Turbine Casing Pipe Forces. Piping connected to the turbine casing may be cold sprung into position so that, when high-temperature steam

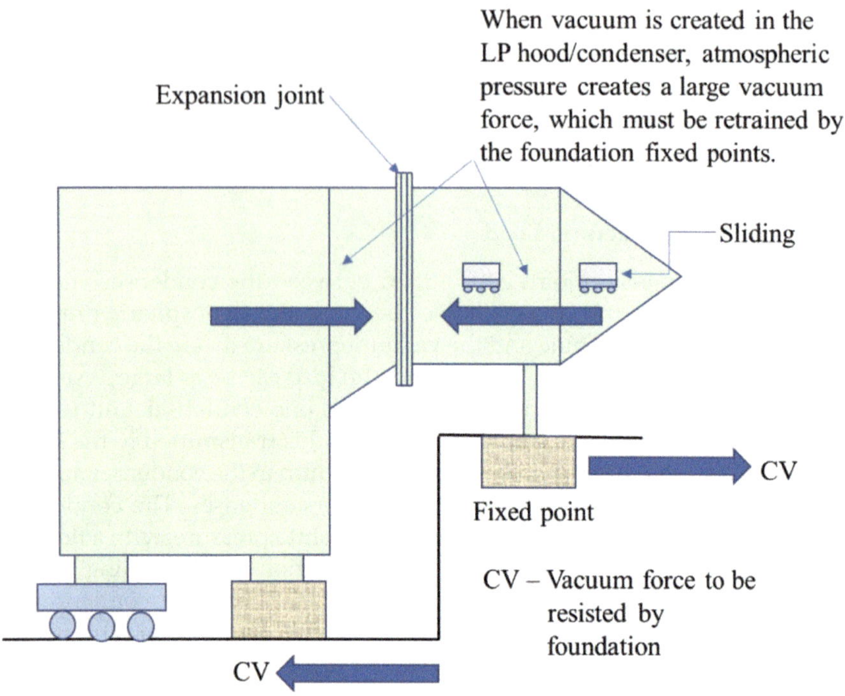

Fig. 4-8. LP hood—condenser vacuum load

heats the system, the stresses will be relieved. Loads created by the connections should be investigated to determine whether they are significant. This may be accomplished by applying the maximum permissible forces and moments that the TG manufacturer specifies. The maximum allowable loads are specified by the TG manufacturer to prevent distortion or overturning of the turbine components. The turbine casing may be assumed to be rigid and the forces then calculated at the support points on the foundation. The types of piping that generate the majority of the loads are

- Main steam inlet piping;
- Reheat steam piping (hot and cold); and
- Extraction steam piping.

Pipe Loading from Equipment Attached to the Foundation. The TG foundation may support assemblies such as the turbine or control valves, interceptor valve, and turbine steam chest. Positioning and aligning the piping for this equipment create restraining forces. The turbine manufacturer's piping is then fitted and welded and the assembly is anchored to the foundation. The remaining steam inlet pipes are then welded to the assembly inlet connections. Different forces are created by thermal expansion during operation. Erection forces and static and dynamic operating forces should be investigated to determine whether they are significant. One significant loading to consider is the loading caused by quick closing of the stop valve attached to the foundation.

The TG manufacturer's piping forces are used whenever specified. These loads are the maximum allowable loads on the sole plates. The direction and point of application of these loads are provided by the manufacturer. However, if actual loads are known from the plant design discipline, they should be used. The actual load shall not exceed the maximum load specified by the manufacturer.

If the turbine manufacturer does not supply piping forces at the soleplates, the designer should obtain the turbine manufacturer's maximum allowable forces at the pipe connections and then calculate forces at support points to determine whether they are significant enough to be considered in the foundation design.

4.4 EMERGENCY OPERATION LOADS

Emergency operation loads result from an abnormal transient event that may occur more than once, though infrequently, during the life of the turbine generator. Operational integrity of the turbine generator without any damage to the foundation system is expected after such an event.

For turbines, abnormal emergency torque loads can be developed from unexpected operating incidents.

4.4.1 Nonsliding Emergency Load (Nonsliding Standard Load)

Many front- and mid-standards on steam turbines have a sliding interface at one or more locations between the base of the machine and the associated foundation baseplate. This interface allows for thermal movement of the steam turbine. In the unlikely event that this sliding movement is hindered, significant forces can be developed on the steam turbine supports. Similar to thermal friction loads, the nonsliding load will not impart a net resultant load on the foundation, though a nonsliding load can exist across a specific foundation element as described for thermal friction loads. This would be considered an emergency load. Nonsliding loads are provided by the manufacturer.

4.5 CATASTROPHIC EQUIPMENT LOADS

Catastrophic equipment loads result from an abnormal transient event that has a low probability of occurrence during the life of the machine, and for which operational integrity of the turbine generator is not specifically required after occurrence.

The foundation design philosophy for such loads should be that, after an occurrence of catastrophic equipment loads, some inelastic deformation and foundation damage may be tolerated, but not to the level of catastrophic structural failure.

4.5.1 Turbine Loss of Blade

A turbine rotor must be balanced dynamically within practical limits in order to ensure satisfactory turbine operation and cause no adverse effects on the turbine equipment and/or foundation. However, a more severe unbalance can occur while the unit is in operation. This emergency unbalance condition is predicated on the unlikely event that a last-row blade in a low-pressure rotor breaks loose from the rotor. The loss of this blade, which can range in size from about 20 in. (500 mm) to over 40 in. (1,000 mm) in length, could cause a significant unbalance in the rotor; consequently, a large dynamic force would be imposed on the rotor/bearing/foundation system. The magnitude of this unbalance is a function of the blade weight, its center of gravity with respect to the rotor, and the square of the rotational speed of the rotor. This force is transferred to the foundation through the rotor bearing system (Fig. 4-9).

The magnitude and location of the forces caused by loss or breakage of a last-row blade are supplied by the turbine manufacturer. The loading can be provided in the form of a sinusoidal forcing function for a dynamic analysis, or equivalent static loads for strength design of the foundation.

4.5.2 Bowed Rotor

An unusually severe packing rub or water induction can cause a bowed rotor in the steam turbine HP or IP section. The resulting force will rotate with the rotor, with a magnitude that varies as a square of the turbine speed, and a duration that is dependent on the shutdown speed of the turbine.

A bowed rotor can impose large dynamic forces on the TG foundation. The bowed condition of the rotor will create unbalance forces that are transmitted through the machine bearings to the soleplates. The magnitude of the forces will vary with the square of the speed, the weight of the rotor, and the amount of eccentricity in the rotor.

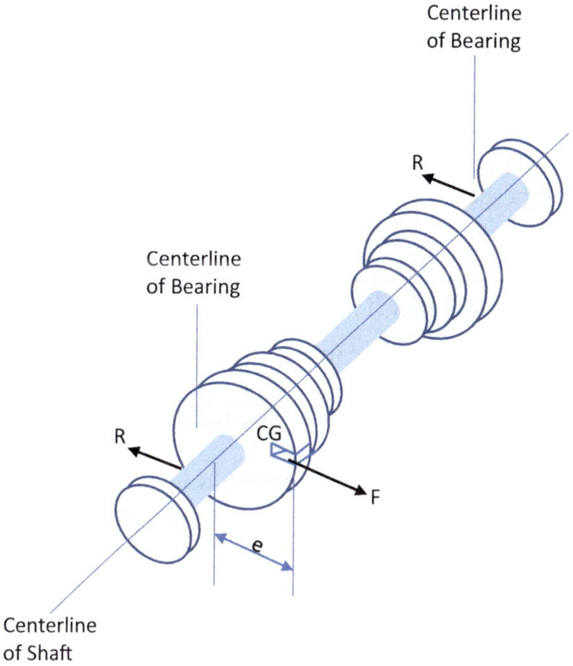

Fig. 4-9. Load due to a missing rotor blade

A bowed rotor can occur in any turbine section and can be the result of the following conditions:

- Unusually severe packing rub;
- Water induction; and/or
- Failure to put the rotor on a turning gear when the machine is shut down.

The first condition induces a differential temperature in the rotor, which causes the rotor to bend, resulting in the unbalance. In the second condition, water induction occurs when a slug of water enters the turbine, causing a differential temperature and bending of the rotor shaft. The third condition is caused by improper operation or system failure and not placing the rotor on turning gear operation while cooling down.

The largest bowed rotor load occurs at the first critical speed for the rotor, which is typically lower than the rated speed. The length of time for the turbine rotor to pass through the critical speed is relatively short while going online. However, the time is much longer when the machine is being taken off line and the rotor coasts through the resonant speed.

The magnitude and location of the forces from a bowed rotor are supplied by the manufacturer of the turbine and will be based on assumptions made by that manufacturer. The loading will be provided in the form of equivalent static loads for the foundation strength design.

Note that some turbine manufacturers assume that a bowed rotor is the worst case of turbine emergency for the high- and intermediate-pressure sections. Other turbine manufacturers assume that the loss of turbine blade (see Section 4.5.1) is the worst turbine emergency and do not consider a bowed rotor emergency load in their foundation design loads for the turbine.

4.5.3 Generator Emergency Torque (Short Circuit Load)

Various short-circuit faults can occur, but a line-to-line short circuit at the generator terminals causes one of the most severe loadings on the TG foundation. Such a fault occurs when any two of the three generator phase terminals are shorted.

The load caused by generator short circuits is provided by the TG manufacturer either as a forcing function for a dynamic analysis, or as equivalent static vertical coupled forces at the soleplate locations where the generator is mounted on the foundation.

Note that some manufacturers treat the short circuit load as an emergency operation load (Section 4.4), and thus suggest a load factor consistent with live load in foundation reinforcement design. The foundation design engineer should consult the manufacturer to understand how this load is estimated, and follow the manufacturer's recommendation as appropriate.

4.6 ENVIRONMENTAL LOADS

Environmental loads are defined by the applicable building codes listed in the civil/structural design criteria document for the project. These loads are specific to a given site location. Turbine manufacturers may provide loads according to site-specific conditions, or maximum loads the TG set can resist, which require scaling down by the design engineer for site-specific conditions.

4.6.1 Seismic Load

For the seismic design of the TG foundation, TG manufacturers usually follow a static force design procedure, and provide machine seismic loads at the supports, assuming that the machine is a nonstructural component (e.g., ASCE 7, Chapter 13). For sites where an appreciable level of seismic activity is possible, the seismic design loads may be significant, especially for the supports that maintain the horizontal alignment of the turbine generator. The design engineer should verify the assumptions associated with the manufacturer-provided seismic loads, as it may be necessary to convert the seismic load that manufacturer applied at the machine centerline to equivalent shears and force-couples at the base plates.

To determine the seismic loads of the TG foundations, the design engineer should follow the applicable building codes. An equivalent lateral force approach is generally used to determine the total seismic base shear for the TG pedestal. See Section 8.3 for additional discussions on seismic design considerations for TG foundations.

4.6.2 Wind Load

Because of the size of concrete TG foundations, for outdoor applications wind loads are not significant in comparison to machine catastrophic loads and/or foundation seismic loads, and therefore usually do not govern the foundation design. However, for CTG configurations with large air inlet duct components, the projected area of the equipment may result in a large wind load. In such cases, the design engineer should calculate the wind load and compare it with the foundation seismic load and/or machine catastrophic loads.

4.6.3 Snow Load

For TG foundations located outdoors, an assumed level of snow accumulation on the turbine generator creates a static load on the turbine generator supports. The design engineer should follow applicable building codes to determine the snow loads, as required. The turbine manufacturer may provide site-specific snow loads for the equipment. The design engineer should work with the TG manufacturer to obtain such loads.

4.7 INSTALLATION AND MAINTENANCE LOADS

Installation and maintenance loads are not concurrent with machine operational loads, but should be considered in the foundation design by the design engineer.

4.7.1 Laydown Load

Certain parts of the turbine generator and associated equipment such as enclosure roof panels may need to be temporarily stored on the turbine generator foundation during periods of maintenance on the turbine generator. The foundation designer should consider the potential for such laydown loads in the design of the turbine generator foundation.

For foundation design purposes, a 250 psf to 400 psf laydown load should be considered on top of an elevated tabletop, as well as on top of the foundation basemat, depending on equipment maintenance weights and associated miscellaneous live loads.

4.7.2 Jacking Post Load

Jacking posts are fabricated or rolled steel shapes that are often embedded in the foundation, against which a jacking device such as a hydraulic cylinder can be placed to move the turbine generator horizontally during installation and maintenance realignments (also see Section 9.2.3). The jacking post design load is dependent on the turbine generator weight, and is provided by the TG manufacturer. This is not an operating load affecting the overall performance of the foundation, but a local maintenance load affecting the area of the embedment (also see Section 9.3.3).

4.7.3 Rotor Removal Load

Temporary loads may be placed on the foundation from jacking devices or other tools used to remove turbine-generator rotors. An example is the jacking devices that are placed underneath the gas turbine casing to prevent ovalization of the lower-half casing when the upper-half casing is removed. This is not an operating load affecting the overall performance of the foundation, but a local maintenance load affecting only the immediate area of its application.

4.8 CONCRETE SHRINKAGE AND CREEP EFFECTS

In reinforced concrete foundations, the time-dependent deflection at bearing locations caused by creep and shrinkage can be two or more times the elastic deformation upon application of the load. However, these additional deflections are generally not considered in the design for the following reasons:

- The initial machine alignment may occur up to 2 years after foundation construction. By this time, most of the basic creep and shrinkage have already taken place. Also, the rate of creep and shrinkage deformation at the time of alignment is well under the initial rate; with a 3- to 5-year period between successive shaft alignments, the incremental values are insignificant.

- The foundation proportions are such that, regardless of the absolute values of the deflections, the differential deflections between bearings are small.

Successful past performance of TG foundations provides ample validity to the previous arguments. Therefore, it is acceptable not to consider the concrete shrinkage and creep effects, unless the machine manufacturer requires such effects to be included in the foundation design.

4.9 LOAD COMBINATION CONSIDERATIONS

The foundation designer is responsible for the selection of the load combinations for the design of the turbine generator foundation based on the applicable building codes listed in the civil/structural design criteria document for the project. To account for foundation loads unique to turbine generators, the following are recommendations for developing load combinations for the turbine generator foundation strength design. The appropriate loads and load combinations for serviceability checks are described in Chapter 7.

4.9.1 Foundation, Machine, and Attachment Self-Weights

Foundation, machine, and attachment self-weights are treated as dead loads per applicable design codes.

4.9.2 Normal Operation Loads

Normal operation loads are defined in Section 4.3. Normal operation loads should be treated as live loads and included in all load combinations that include live loads, with the following additional notes and exceptions:

1. Normal operation loads in some cases can reduce the overall design load. In such cases, it may be appropriate to exclude such loads from certain load combinations. For example, the gas turbine axial thrust load may counteract the seismic force acting in a certain direction, and thus should be excluded from that load combination.
2. Normal operation loads should be included in the load combinations for turbine generator emergency operation or catastrophic loads, except when some normal operation loads are superseded by the specific emergency operation or catastrophic load (see also items 5 and 6).
3. In some cases, building codes may allow a reduced load factor for certain types of live loads. For turbine generator normal operational

loads, the most conservative load factor applicable to live loads, that is, the largest in magnitude, should be used.

4. Since thermal friction loads are transitory, thermal friction loads are normally not included in load combinations for low-probability, transient events such as seismic loads, turbine generator emergency loads, and turbine generator catastrophic loads. On the other hand, since thermal expansion loads are not transitory, they should be considered in all such load combinations.
5. The normal operational torque load is superseded by emergency torque load, and therefore the two are not included in the same load combinations.
6. The gas turbine axial thrust pressure differential foundation load may be specified at two load states: normal operating conditions and an emergency shutdown level. Since the latter supersedes the former, the two should not be included in the same load combinations.
7. Since the magnitude of steam turbine condenser vacuum load can be accurately predicted, the vacuum loads should be treated as dead load.

4.9.3 Emergency Operation Loads

Emergency operation loads are defined in Section 4.4. Unique load combinations need to be developed for emergency operation loads. Emergency operation loads should be treated as live loads, with the following considerations:

1. When developing load combinations for emergency operation loads, it is not necessary to consider emergency operation loads acting concurrently with seismic, wind, or other emergency loads and machine catastrophic loads.
2. When developing load combinations for emergency operation loads, it is assumed that the emergency operation loads act concurrently with the turbine generator normal operation loads, except in those cases where the emergency operation load supersedes the normal operation load. For example, the emergency torque foundation load supersedes the normal operation torque load, and therefore the two loads should not be included in the same load combination.

4.9.4 Catastrophic Equipment Loads

Catastrophic loads are defined in Section 4.5. Unique load combinations need to be developed for catastrophic loads, with the following considerations:

1. Catastrophic loads are assigned a load factor of 1.0, unless specifically required or suggested by the TG manufacturer.
2. Catastrophic loads need not be combined with seismic or wind loads.
3. For steam turbine dual-flow LP exhaust hoods, it is assumed that the loss-of-blade load acts at either end of the exhaust hood, but not at both ends simultaneously.
4. It is assumed that the gas turbine loss of compressor blade load and loss of turbine blade load do not act simultaneously.
5. Although generator short-circuit loads are accidental, unlike other accidental loads, some TG manufacturers expect the machine and foundation to be fully functional after such a load, and thus may recommend a load factor of 1.6 that is consistent with live loads, in lieu of a load factor of 1.0 associated with a catastrophic type loading scenario (also see Sections 4.4 and 4.5.3).

4.9.5 Environmental Loads

Environmental loads are defined in Section 4.6. For TG foundations, snow and wind loads are usually not significant in comparison to seismic loads, unless the foundation is to be designed for extreme wind or snow environments for a particular project. Load combinations for seismic loads are dictated by the project-specific building code. Seismic loads should be considered as follows:

1. The seismic loads resulting from the horizontal seismic design acceleration should be assumed in two directions (one aligning with and one perpendicular to the turbine generator shaft centerline axis). The orthogonal combination effects should be considered per project-specific building code requirements.
2. The seismic loads resulting from the vertical seismic design acceleration should be assumed to act concurrently with the seismic loads resulting from the horizontal seismic design acceleration, unless the project-specific building code grants relief based on the level of seismic risk.

4.9.6 Installation and Maintenance Loads

Installation and maintenance loads are defined in Section 4.7. These loads should be treated as live loads in applicable load combinations.

4.9.7 Example Load Combinations for Turbine Generator Foundation Loads

The following is an example for developing load combinations to account for the foundation loads that are unique to a turbine generator foundation.

Table 4-1. ASCE 7 Load Combinations

1.4D
1.2D + 1.6L + 0.5(L_r or S or R)
1.2D + 1.6(L_r or S or R) + (L or 0.5W)
1.2D + 1.0W + L + 0.5(L_r or S or R)
1.2D + 1.0E + L + 0.2S
0.9D + 1.0W
0.9D + 1.0E

Note: L = Live; W = Wind; E = Seismic; S = Snow; and L_r = Roof Live Load.
Source: ASCE 7.

This example is based on a strength design approach, as stated in ASCE 7. It is also based upon a strength-level limit for wind and earthquake loads rather than an equivalent wind and earthquake loading used with allowable stress design (see Tables 4-1 and 4-2).

These load combinations are not intended for static deflection or dynamic analysis, or for soil pressures or pile reactions. Load cases may need to be expanded for reversing emergency loads, for catastrophic forces acting horizontally and vertically, and loss-of-blade (LBT) or loss-of-compressor-blade (LBC) acting separately (for a gas turbine). For design of embedments such as anchor bolts, the load combinations using 0.9D when dead load opposes the effects of applied machine loads are critical. However, for the design of members in which the anchor bolts are embedded, the load combinations using 1.2D when dead load is additive to the effects of machine loads are critical. For single-shaft applications, catastrophic loads for generators, gas turbines, and steam turbines need not be applied simultaneously.

In addition to the load combinations per applicable building code(s), specific load combinations recommended by machine manufacturers, which may be different from or supplement to code requirements, should be considered in the foundation design.

4.9.7.1 Example Load Combinations Developed for Gas Turbine.

A— Non-Operating Loads

1.4D
1.2D + 1.6 (jacking post, maintenance, laydown, rotor removal)

B— Normal Operation (including Environmental)

1.2D + 1.6(TF + NTQ + NTH + ATH + L) + 0.5S
1.2D + 1.0(TF + NTQ + NTH + ATH + L) + 1.6S
1.2D + 0.5W + 1.6S

Table 4-2. Load Key and Applicability

Type	Gas Turbine	Steam Turbine	Generator
Non-operating	L, D, Jacking Post, Maintenance, Laydown, Rotor Removal		
Environmental	W, S, E		
Normal operation	D, TF, NTQ, NTH	D, TF, TE, NTQ, CV, ATH, CO	D, TF, NTQ
Emergency	ETH	NSS	
Catastrophic	LBC, LBT	BR, LBT	ETQ

Key:
D: Dead
L: Live
W: Wind
E: Earthquake
S: Snow
CV: Condenser Vacuum
TF: Thermal Friction
TE: Thermal Expansion
NTQ: Normal Operating Torque
NTH: Normal Thrust / Normal Pressure Differential
ATH: Axial Thrust / Flex Coupling
ETQ: Emergency Torque (Short Circuit)
ETH: Axial Emergency Thrust / Shutdown Pressure Differential
LBC: Loss of Compressor Blade For Gas Turbine
LBT: Loss of Blade
BR: Bowed Rotor
NSS: Nonsliding Standard (Nonsliding Emergency Load)
CO: Condenser Operation
COTF: Condenser Operation Thermal Friction

Notes:
1. Piping/valve trip loads not considered.
2. Earthquake and wind are strength-level loads
3. The magnitudes of live load should be appropriately reduced during normal operation.
4. W and S are applicable depending on exposure of equipment.
5. Special seismic loading combinations as required by ASCE 7 not shown.

Source: Courtesy of GE Power; reproduced with permission.

1.2D + 1.0E + 1.0(TF + NTQ + NTH + ATH + L) + 0.2S
1.2D + 1.0W + 1.0(TF + NTQ + NTH + ATH + L) + 0.5S
0.9D + 1.0W
0.9D + 1.0E

C— Emergency Operation

νD + 1.6(TF + NTH + ATH + L) + 0.5S
νD + 1.0(TF + NTQ + ETH + ATH + L) + 1.6S

where ν = 1.2 when D is additive to ETH, and ν = 0.9 when D counteracts the effects of ETH.

D— Catastrophic Event
D + (LBC or LBT or ETQ) + (TF + NTQ + NTH + ATH + L) + S

Note: Some machine manufacturers may consider ETQ an emergency operational event.

4.9.7.2 Example Load Combinations Developed for Steam Turbine.
A- Non-Operating
1.4D + (jacking post, maintenance, laydown, rotor removal)

B— Normal Operation (including Environmental)

1.2(D + CV) + 1.6(TF + TE + NTQ + ATH + L + CO + COTF) + 0.5S
1.2(D + CV) + 1.0(TF + TE + NTQ + ATH + L + CO + COTF) + 1.6S
1.2(D + CV) + 0.5W + 1.6S
1.2(D + CV) + 1.0W + 1.0(TF + TE + NTQ + ATH + L + CO + COTF) + 0.2S
1.2(D + CV) + 1.0E + 1.0(TF + TE + NTQ + ATH + L+ CO + COTF) + 0.5S
0.9(D + CV) + 1.0E
0.9(D + CV) + 1.0W

C— Emergency Operation

ν(D + CV) + 1.6(TF + TE + ATH + L + CO) + 0.5S
ν(D + CV) + 1.0(TF + TE + NTQ + NSS + L+ CO) + 1.6S

where ν = 1.2 when D is additive to ETH, and ν = 0.9 when D counteracts the effects of ETH.

D— Catastrophic Event
(D + CV) + (BR or LBT or ETQ) + (TF + TE + NTQ + NTH + ATH + L) + S

Note: Some machine manufacturers may consider ETQ an emergency operational event.

4.9.7.3 Example Load Combinations Developed for Generator.
A— Non-Operating
1.4D + (jacking post, maintenance, laydown, rotor removal)

B— Normal Operation (including Environmental)

1.2D + 1.6(TF + NTQ + L) + 0.5S
1.2D + 1.0(TF + NTQ + L) + 1.6S
1.2D + 0.5W + 1.6S
1.2D + 1.0W + 1.0(TF + NTQ + L) + 0.5S
1.2D + 1.0E + 1.0(TF + NTQ + L) + 0.2S
0.9D + 1.0W
0.9D + 1.0E

C— Emergency Operation

νD + 1.0(TF + ETQ + L) + 0.5S
νD + 1.0(TF + ETQ + L) + 1.6S

where ν = 1.2 when D is additive to TF and L, and ν = 0.9 when D counteracts the effects of TF and L.

D— Catastrophic Event
D + (LBC or LBT or ETQ) + (TF + NTQ + NTH + L) + S

Note: Some machine manufacturers may consider ETQ an emergency operational event.

REFERENCES

ACI (American Concrete Institute). (2008). "Code requirements for reinforced concrete chimneys." *ACI 307-08*, Farmington Hills, MI.
ACI. (2007). "Reinforced concrete design for thermal effects on nuclear power plant structures." *ACI 349.1R-07*, Farmington Hills, MI.
ASCE. (2010). "Minimum design loads for buildings and other structures." *ASCE/SEI 7-10*, Reston, VA.
General Electric. (2015). *Turbine generator foundations. GEK 63383e.*
International Code Council. (2012). "International Building Code." *IBC-2012*, Country Club Hills, IL.
ISO (International Organization for Standardization). (2003). "Mechanical vibration: balance quality requirements for rotors in a constant (rigid) state, Part 1: Specification and verification of balance tolerances." 2nd Ed., *ISO 1940-1*, Geneva, Switzerland.

CHAPTER 5

MODELING OF THE SOIL AND PILE RESPONSE TO DYNAMIC LOADS

5.1 INTRODUCTION

In practice, subgrade and pile impedances are used for modeling the soil and pile response of a given foundation under applied loads. For instance, the response of a rigid foundation to static or dynamic load (P) arises solely from the deformation (U) of the supporting soil. Therefore, the static soil stiffness K, defined as $K = P/U$, is the only parameter needed to model the soil response to static loads acting on a rigid foundation. In an analogous manner, the dynamic impedance of the subgrade $K(\Omega)$, defined as $K(\Omega) = P(t)/U(t)$, is used to model the soil response of a rigid foundation under dynamic loads (e.g., from machinery operation). In particular, six subgrade dynamic impedances are required—three translational and three rotational—to formulate the dynamic equilibrium equations of a rigid foundation supported on the soil/subgrade surface, such as that shown in Fig. 5-1. Additional impedances (i.e., "cross-coupling" horizontal rocking impedances) are required when the rigid foundation is embedded or pile supported. Furthermore, a large coupled dynamic impedance matrix is required for modeling the soil response of flexible foundations. In any case, these dynamic impedances are a function of the foundation geometry, the soil properties (which are strain-dependent), and the vibration frequency of the machine (f, Ω), which varies during the machine startup and shutdown. In general, subgrade and pile impedances are computed using analytical, numerical, or experimental methods, as discussed later.

In practice, structural engineers are mostly concerned with the use of the subgrade and pile impedances, for modeling the soil behavior and calculating the dynamic response of a given foundation. On the other hand, geotechnical engineers typically face the task of calculating the subject of

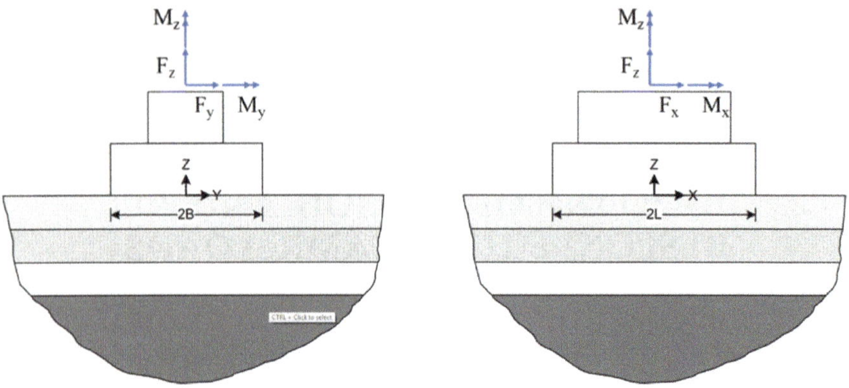

Fig. 5-1. Degrees of freedom of a rigid foundation block resting on a layered half-space

dynamic impedances. While the calculation of dynamic impedances is one of the first and most important steps for the dynamic analysis of a machine foundation, this section begins by showing the use of these dynamic impedances in structural engineering practice. A rigid foundation is used to keep the analysis simple and illustrate the general approach for calculating the dynamic response of machine foundations to dynamic loads, which can be summarized as follows:

1. Determine subgrade or pile impedances for the operating frequency of the machine (Ω_0), and any other frequencies (Ω) of interest during the machine startup and shutdown. The procedures used for these calculations are discussed later.
2. Determine the forcing functions acting on the machine foundation, as described in Chapter 4.
3. Use Newton's second law and elasticity theory to formulate a mathematical model of the soil-foundation system. In practice, this step is typically done using specialized finite element codes as described in Chapter 6.
4. Solve the system of equations resulting from step 3, for each frequency of interest. This step is typically done by the computer code used for modeling the foundation.
5. Report the displacement, velocity, and acceleration amplitudes at locations of interest, for instance, at the machine supports and/or bearing locations.

The design decisions associated with the results obtained from the calculations described here are discussed in Chapter 7, which provides detailed serviceability requirements and acceptance criteria. Additionally, it must be noted that step 4 here requires the selection of an adequate dynamic analysis method, whether in the time domain or in the frequency domain. For instance, modal analysis can cause errors when the impedance functions are strongly frequency dependent. However, this is not the case if the direct integration method in the time domain or the frequency response method in the frequency domain are used.

5.1.1 Vertical Response of a Rigid Foundation

Consider the foundation shown in Fig. 5-2, which has a radius R, thickness t, and a foundation mass m_f. The operating frequency of the machine is f_0 ($\Omega_o = 2\pi f_0$), the machine mass is m_m, the rotating mass is m_r, and the mass eccentricity is e_r, which results in an unbalanced load $F_z(t)$, as described in Chapter 4. On the other hand, the dynamic impedance of the layered subgrade soil for vibration in the vertical direction is $K_z(\Omega)$, where it must be noted that the subgrade impedance is typically provided by the geotechnical engineer as a function of the machine frequency ($\Omega = 2\pi f$). Additionally, it is important to emphasize that $\Omega = 2\pi f$ is different from $\Omega_o = 2\pi f_o$; the former corresponds to the vibration frequency of the machine during startup or shutdown, while the latter corresponds to the operating frequency of the machine.

The generic foundation shown in Fig. 5-2 (Gazetas 1991a) is used to illustrate the degrees of freedom (i.e., vibration mode) considered while deriving the equation of motion. For this simple foundation, the horizontal

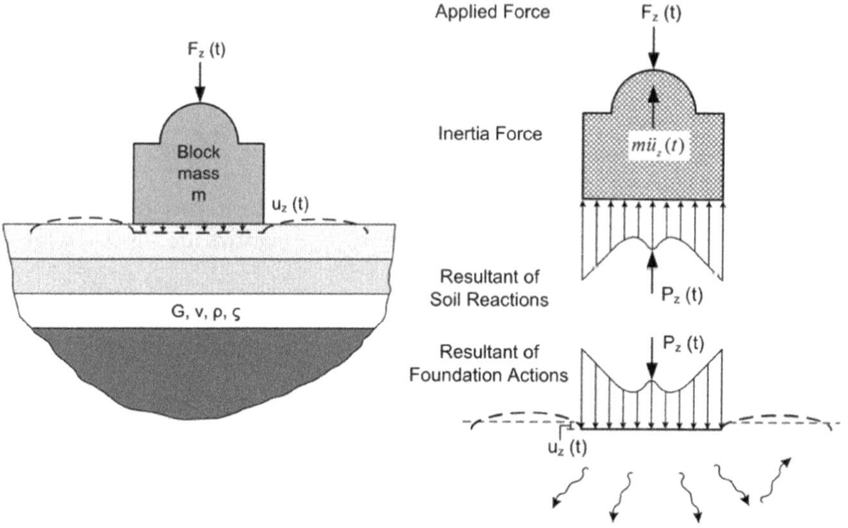

Fig. 5-2. Analysis of dynamic equilibrium of a vertically vibrating foundation block
Source: Adapted from Gazetas (1991b).

eccentricity between the center of gravity (CG) of the machine-foundation system and the center of soil pressures or resistance (CR) is assumed to be zero; otherwise coupling between rocking and vertical vibration will need to be considered. Furthermore, the foundation is under a vertical harmonic force $F_z(t)$ acting along the CG; therefore the only displacement experienced by the foundation is $u_z(t)$.

A detailed derivation of the equation of motion for the foundation depicted in Fig. 5-2 can be found elsewhere (e.g., Gazetas 1991a). In summary, the dynamic equilibrium equation of the block foundation is

$$P_z(t) + m\ddot{u}_z(t) = F_z(t) \tag{5-1}$$

The soil resistance, acting at the CR, is provided by $P_z(\Omega) = K_z(\Omega)u_z(t)$, where $K_z(\Omega)$ is the dynamic impedance of the subgrade for vertical vibration, which can be modeled as a frequency-dependent spring and dashpot as follows: $K_z(\Omega) = \bar{k}_z(\Omega) + i\Omega C_z(\Omega)$, where $i = \sqrt{-1}$. Therefore, the equation of motion can be rewritten as follows:

$$\begin{aligned} &K_z(\Omega)u_z(t) + m\ddot{u}_z(t) = F_z(t) \\ &(\bar{k}_z(\Omega) + i\omega C_z(\Omega))u_z(t) + m\ddot{u}_z(t) = F_z(t) \end{aligned} \tag{5-2}$$

Eq. (5-2) can be transformed into an algebraic equation, given that the loading from the rotating machine is harmonic ($F_z(t) = F_0 e^{i\Omega t}$), and therefore the response of the foundation is harmonic as well ($u_z(t) = u_0 e^{i(\Omega t + \varphi)}$). In particular, after $F_z(t)$ and $u_z(t)$ are replaced in Eq. (5-2), the displacement of the foundation can be obtained by algebraic manipulation. It must be noted that $u_z(\Omega)$ and $K_z(\Omega)$ are complex functions of the rotation frequency of the machine ($\Omega = 2\pi f$):

$$u_z(\Omega) = \frac{F_0}{K_z(\Omega) - \Omega^2 m} \tag{5-3}$$

Eq. (5-3) can be evaluated using complex algebra procedures or standard computer codes for mathematical computations, as illustrated in Appendix C, Section C.1.

5.1.2 Horizontal and Rocking Response of a Rigid Foundation

Fig. 5-3 is used to illustrate the degrees of freedom and forces considered while deriving the equation of motion of a generic foundation under harmonic lateral load. For this type of loading, the horizontal and rocking vibrations of the soil-foundation-machine system are coupled, since the lines of action of the different horizontal forces are offset, as shown in Fig. 5-3. Therefore, the mass (m) and mass moment of inertia about the x-axis (I_{0x}) of the combined machine-foundation system are required to formulate the motion equation. The horizontal unbalanced load $F_y(t)$ pro-

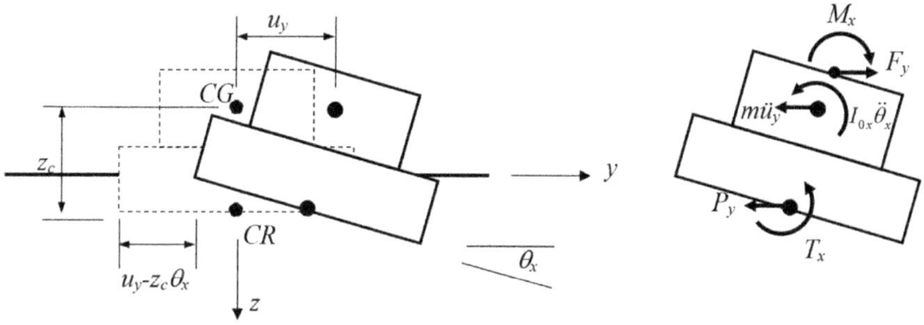

Fig. 5-3. Analysis of dynamic equilibrium of a horizontally vibrating foundation block
Source: Adapted from Gazetas (1991b).

duces a rocking moment $M_x(t) = F_y(t)z_r$, where z_r is the vertical distance from the CG of the machine-foundation system to the point of application of horizontal load (i.e., the rotor location).

A detailed derivation of the equation of motion for the foundation depicted in Fig. 5-3 can be found elsewhere (e.g., Gazetas 1991a). In summary, the dynamic equilibrium equation of the block foundation is given by Eqs. (5-4) and (5-5), where the summation of forces and moments is done with respect to the foundation CG. The soil resistance acts at the center of resistance (CR) and is provided by $P_y(t)$ and $T_x(t)$, which are given by Eqs. (5-6) and (5-7), where K_y, K_{rx}, K_{yrx} are respectively the horizontal, rocking, and cross-coupling horizontal-rocking dynamic impedances of the subgrade supporting the foundation. Therefore, after substitution of Eqs. (5-6) and (5-7) in (5-4) and (5-5), the equation of motion can be rewritten as shown in Eqs. (5-8) and (5-9), where the time variable has been dropped for the sake of brevity and clarity.

$$P_y(t) + m\ddot{u}_y(t) = F_y(t) \tag{5-4}$$

$$T_x(t) - P_y(t)\cdot z_c + I_{0x}\ddot{\theta}_x(t) = M_x(t) \tag{5-5}$$

$$P_y(t) = K_y(u_y(t) - z_c\theta_x(t)) + K_{yrx}\theta_x(t) \tag{5-6}$$

$$T_x(t) = K_{rx}\theta_x(t) + K_{yrx}(u_y(t) - z_c\theta_x(t)) \tag{5-7}$$

$$K_y(u_y - z_c\theta_x) + K_{yrx}\theta_x - \Omega^2 m u_y = F_y \tag{5-8}$$

$$K_{rx}\theta_x + K_{yrx}(u_y - z_c\theta_x) - [K_y(u_y - z_c\theta_x) + K_{yrx}\theta_x]z_c - \Omega^2 I_{0x}\theta_x = M_x \tag{5-9}$$

Given that the dynamic impedances are complex-valued functions of the vibration frequency, these simultaneous equations can be solved for harmonic loading using complex algebra procedures or standard computer codes for mathematical computations. Such procedures are illustrated in Appendix C, Section C.2.

5.1.3 Chapter Scope

The remainder of this chapter is devoted to describing the main procedures used for calculating the dynamic impedance of subgrade and pile-supported foundations. In practice, these impedances are calculated using specialized computer codes in all but the simplest cases. In its simplest form, a dynamic impedance is provided as a frequency dependent complex function ($K(\Omega)=\bar{k}(\Omega)+i\Omega C(\Omega)$) or as a spring-dashpot system, with the *spring* becoming negative for certain frequencies; therefore, often this "spring" cannot be directly used in standard structural analysis computer codes. Thus, the practical use and application of dynamic impedances require a clear understanding of the theory behind their calculation, along with several principles of soil dynamics, which are not covered in the regular curriculum of structural engineering programs. This chapter aims to fill this gap by providing structural engineers with the basic tools for understanding the calculation and use of foundation impedance functions. Conversely, this chapter does not aim to provide detailed procedures for calculation of foundation dynamic impedances other than for the simplest/benchmark cases, which can be used for validation of the specialized computer codes used in practice for the calculation of the subject impedances.

5.2 DYNAMIC IMPEDANCE DEFINITION

For a linear dynamic system excited by a harmonic force (or moment) $P(t) = P_0 e^{i\Omega t}$, the response (e.g., displacement or rotation) is known to be harmonic $U(t) = u_0 e^{i(\Omega t-\varphi)}$. By definition [see Eq. (5-10)], the dynamic impedance of such a system is the ratio of the load $P(t)$ acting on the system to its corresponding displacement response $U(t)$; therefore, it has stiffness units (i.e., force per unit length or torque per radian). The displacement response depends on the forcing frequency (Ω) and typically lags the loading; in other words, the load and the response are not in phase. Therefore, for mathematical convenience, traditionally complex notation is used to represent the dynamic impedance of a foundation, given that complex numbers encapsulate magnitude and phase information in a single entity/number, as shown in Eq. (5-10).

$$K(\Omega)=\frac{P(t)}{u(t)}=\frac{P_0 e^{i\Omega t}}{u_0 e^{i(\Omega t-\phi)}}=\frac{P_0}{u_0}e^{i\phi} \qquad (5\text{-}10)$$

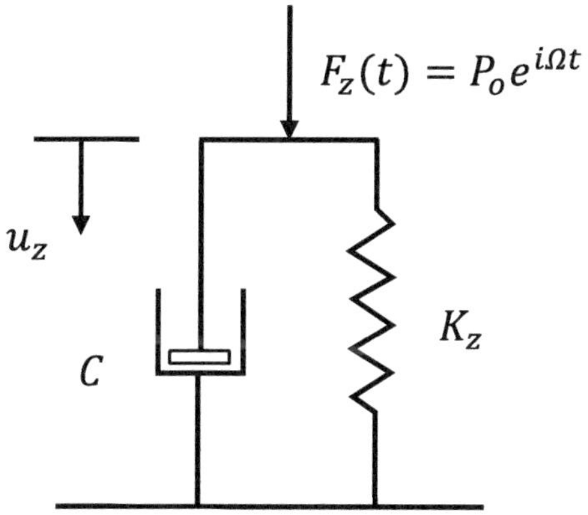

Fig. 5-4. Spring dashpot system
Source: Adapted from Lysmer (1965).

In most soil dynamics publications, dynamic impedances are expressed per Eq. (5-11), as originally proposed by Lysmer (1965) (see Fig. 5-4); where both $\bar{K}$ and C are functions of the forcing frequency Ω. The real component, $\bar{K}$, is called the *dynamic stiffness* and reflects the stiffness and inertia of the supporting soil. Since soil properties are practically frequency independent, the variation of the *dynamic stiffness* with frequency results from the inertial response of the soil, as shown in Section 5.2.2 for a single degree of freedom (SDOF) system. The imaginary component, ΩC, is expressed as the product of the circular frequency Ω times the "dashpot coefficient," C. The latter reflects the hysteric/material damping of the soil and the radiation damping from energy carried by waves spreading away from the foundation. Eq. (5-11) is widely used in soil dynamics practice because it allows an analogy between the actual foundation-soil system and an equivalent system consisting of the same foundation, supported by an analogue *spring* and *dashpot* with properties equal to $\bar{K}$ and C, respectively.

$$K(\Omega) = \bar{K} + i\Omega C \tag{5-11}$$

In the previous discussion, it must be emphasized that the loading and response (vibration amplitude) of a machine foundation are always real; complex numbers are used just for mathematical convenience given that they greatly simplify calculations involving harmonic functions, such as the loading from rotating machines. Before dealing with the dynamic impedance of soils and piles, the definition is clarified in the following sections by calculating the dynamic impedance of simple mechanical systems of practical importance for the dynamic analysis of machine foundations.

5.2.1 Dynamic Impedance of a Spring Dashpot System

Consider the system shown in Fig. 5-4, which is of interest since its dynamic displacement function closely matches that of a massless foundation resting on homogeneous soil, as shown by Lysmer (1965). This system has only one degree of freedom (DOF) and its motion equation is as follows:

$$C\dot{u}_z + K_z u_z = P_o e^{i\Omega t} \tag{5-12}$$

Substitution of the displacement response $u_z(t) = u_o e^{i(\Omega t-\varphi)}$ and algebraic manipulations yield the dynamic impedance given in Eq. (5-14).

$$(K_z + C\Omega i)u_o e^{i(\Omega t-\varphi)} = P_o e^{i\Omega t} \tag{5-13}$$

$$K(\Omega) = \frac{F_z(t)}{u_z(t)} = \frac{P_o e^{i\Omega t}}{u_o e^{i(\Omega t-\phi)}} = K_z + C\Omega i \tag{5-14}$$

Here the real part is equal to the static stiffness of the system and the imaginary part is frequency-dependent and related to the damping of the system. Lysmer (1965) shows that the vertical displacement function of this system closely matches that of a massless foundation resting on homogeneous soil. The subject displacement function can be obtained by algebraic manipulation of Eq. (5-13) as illustrated next:

$$u_o e^{-i\phi} = \bar{u}_o = \frac{P_o}{K_z + C\Omega i} = \frac{P_o}{K_z}\frac{1}{1+\frac{C}{K_z}\Omega i} = \frac{P_o}{K_z}\frac{1}{1+a_0 i} = \frac{P_o}{K_z}F \tag{5-15}$$

$$F = \frac{1}{1+a_0 i} \qquad a_0 = \frac{C}{K_z}\Omega \tag{5-16}$$

$$F_1 = \frac{1}{1+a_0^2} \tag{5-17}$$

$$F_2 = \frac{-a_0}{1+a_0^2} \tag{5-18}$$

In these equations, F is called the displacement function and a_0 is a dimensionless parameter called the dimensionless frequency ratio. The real and imaginary parts of Eq. (5-16) are respectively given by Eqs. (5-17) and (5-18), which are plotted in Fig. 5-5. Lysmer (1965) showed that such displacement functions closely match those of a rigid massless foundation resting on the surface of a homogeneous soil profile (elastic half-

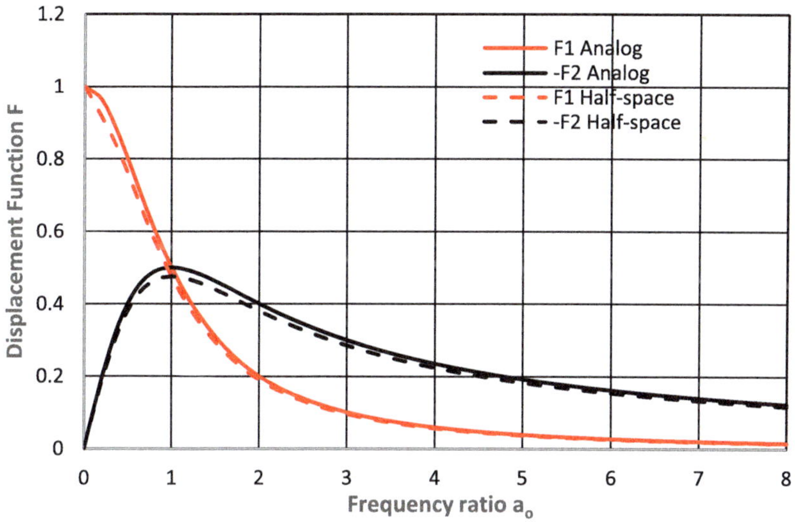

Fig. 5-5. Comparison of vertical displacement functions for a spring dashpot system ($K_1 = 1$, $C_1 = 0.85$) and a rigid footing on uniform soil
Source: Adapted from Lysmer (1965).

space). Such similarity was used by Lysmer (1965) to introduce the idea that the vertical response of a rigid footing can be represented by a SDOF system with constant stiffness and damping, in other words frequency independent subgrade parameters. This simplified approach is typically referred to as Lysmer's analogy and was further extended to other vibration modes, using fictitious masses, as discussed by Richart et al. (1970). However, the range of applications for the subject frequency independent analogy is rather limited and typically not used in practical applications, particularly for high-frequency machines such as turbine generators.

5.2.2 Dynamic Impedance of a Single Degree of Freedom System

Consider the SDOF system shown in Fig. 5-6, which is of interest since it can be used to demonstrate that a linear dynamic system, such as a linear elastic subgrade, can be modeled as a frequency dependent *dynamic spring* and dashpot system. Furthermore, the dynamic impedance of a SDOF is used to formulate the important concept of *added mass*, which in practice is used for modeling negative dynamic *soil springs*. The equation of motion for this dynamic system is as follows:

$$m\ddot{u}_z + C\dot{u}_z + K_z u_z = P_0 e^{i\Omega t} \tag{5-19}$$

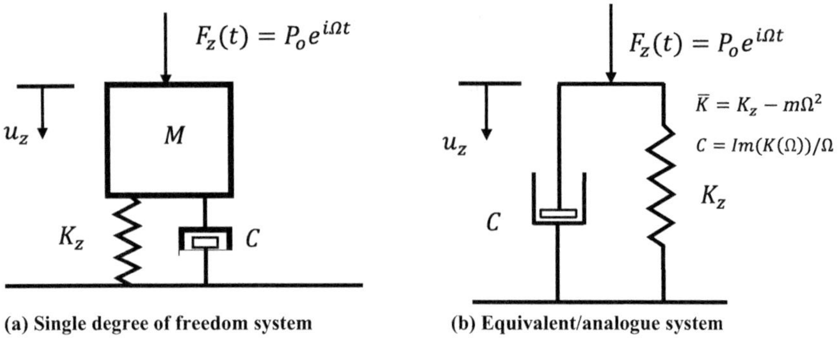

Fig. 5-6. Single degree of freedom system under a vertical forcing function; (a) single degree of freedom system; (b) equivalent analogue system

Substitution of the displacement $U(t) = u_0 e^{i(\Omega t - \varphi)}$ and algebraic manipulations yield the dynamic impedance of the system given by Eq. (5-21), with the real part $Re(K(\Omega)) = \bar{K}(\Omega) = K_z - m\Omega^2$ and the imaginary part $Im(K(\Omega)) = C\Omega$. Similarly, the impedance magnitude and phase angle are respectively given by Eqs. (5-22) and (5-23). As already mentioned, the dynamic impedance has a unit of stiffness (force per unit length); therefore, in practice the real part is associated/modeled by a *spring* (or dynamic stiffness). Similarly, the imaginary part is normalized by the frequency of the forcing function, and associated to a viscous dashpot $C = Im(K(\Omega))/\Omega$, as shown in Fig. 5-6(b). Nevertheless, it is very important to notice that the so-called spring is made of the actual static stiffness of the system (K_z) and the system inertia ($m\Omega^2$). Therefore, this special spring has very interesting characteristics, as discussed next.

$$(K_z + C\Omega i - m\Omega^2) u_0 e^{i(\Omega t - \varphi)} = P_0 e^{i\Omega t} \tag{5-20}$$

$$K(\Omega) = \frac{F_z(t)}{u_z(t)} = \frac{P_0 e^{i\Omega t}}{u_0 e^{i(\Omega t - \phi)}} = |K(\Omega)| e^{i\phi} = K_z - m\Omega^2 + C\Omega i \tag{5-21}$$

$$|K(\Omega)| = \sqrt{(K_z - m\Omega^2)^2 + (C\Omega)^2} \tag{5-22}$$

$$\tan\phi = \frac{C\Omega}{K_z - m\Omega^2} \tag{5-23}$$

Following the previous discussion, it is instructive to study the behavior of the dynamic impedance of the subject SDOF system. For this purpose, Eq. (5-21) is normalized by the static stiffness K_z as shown in Eq. (5-24),

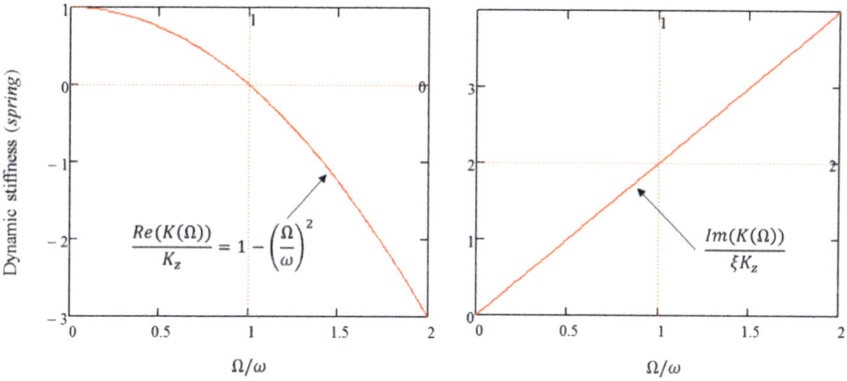

Fig. 5-7. Single degree of freedom system under a vertical forcing function

where $\omega = \sqrt{K_z/m}$ is the natural frequency of the SDOF system and $\xi = C/2\sqrt{K_z m}$ is the corresponding damping ratio. The real and imaginary parts of Eq. (5-24) are plotted in Fig. 5-7.

$$\bar{k} = \frac{K(\Omega)}{K_z} = 1-\left(\frac{\Omega}{\omega}\right)^2 + 2\xi\left(\frac{\Omega}{\omega}\right)i \tag{5-24}$$

As can be seen in Fig. 5-7, the imaginary part is linearly proportional to the excitation frequency, while the real part has a parabolic variation and becomes zero for $\omega = \Omega$, when the system reaches/approaches the resonant condition. At resonance, the system experiences the maximum displacement response and therefore the dynamic impedance magnitude reaches a minimum as illustrated in Fig. 5-8(a). In fact, at resonance most of the forcing/unbalance force is resisted by the system damping. Additionally, for forcing frequencies (Ω) greater than the natural frequency of the system (i.e., for $\Omega/\omega > 1$), the dynamic stiffness term becomes negative. In other words, for $K_z < m\Omega^2$ the forcing function and the displacement of the mass are out of phase; their motion is in opposite directions, $\phi \sim 180°$, as shown in the Fig. 5-8(b) plot of Eq. (5-23). In other words, for $\Omega < \omega$ the static stiffness term dominates (i.e., $K_z > m\Omega^2$), whereas for $\Omega > \omega$ the inertial term dominates (i.e., $K_z < m\Omega^2$).

5.3 DYNAMIC IMPEDANCE OF RIGID FOUNDATIONS

In general, six dynamic impedances are required—three translational and three rotational—to formulate the dynamic equilibrium equations of a rigid foundation supported on the soil/subgrade surface, such as that

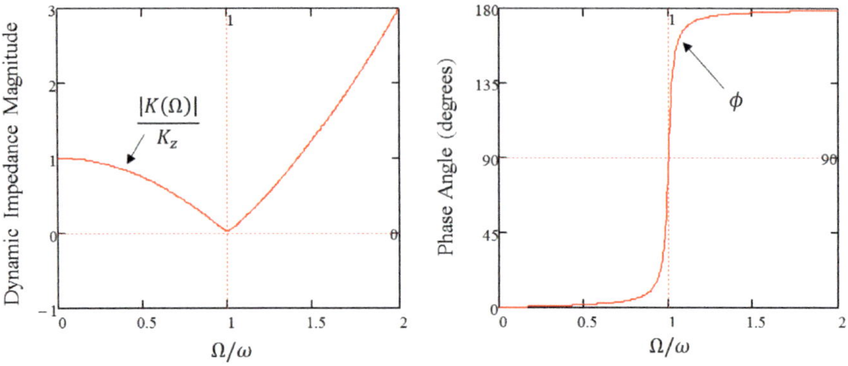

Fig. 5-8. Variation of dynamic impedance magnitude and phase angle (ϕ) of a SDOF system

shown in Fig. 5-1. Additional impedances (i.e., "cross-coupling" horizontal-rocking impedances) are required when the rigid foundation is embedded or pile-supported. For mathematical convenience, these impedances can be represented in matrix form as shown by Eq. (5-25), called the dynamic impedance matrix of a rigid foundation.

$$K(\Omega)=\begin{bmatrix} K_x & 0 & 0 & 0 & K_{xry} & 0 \\ 0 & K_y & 0 & K_{yrx} & 0 & 0 \\ 0 & 0 & K_z & 0 & 0 & 0 \\ 0 & K_{yrx} & 0 & K_{rx} & 0 & 0 \\ K_{xry} & 0 & 0 & 0 & K_{ry} & 0 \\ 0 & 0 & 0 & 0 & 0 & K_{rz} \end{bmatrix} \tag{5-25}$$

Each element (impedance) of this matrix is a function of the foundation geometry (e.g., circular, rectangular, strip), foundation type (soil- versus pile-supported), degree of embedment (e.g., surface, partially embedded, fully embedded), the foundation stiffness (rigid versus flexible), the properties of the soil profile (e.g., homogeneous versus layered), the vibration frequency (Ω) and the vibration mode (e.g., vertical K_z, lateral K_x, K_y, rocking K_{rx}, K_{ry}, or torsion K_{rz}). Given their massive dimensions, block foundations are typically approximated as rigid, which simplifies their analytical treatment. Conversely, typical STG foundations cannot be modeled as rigid, given the high operation frequency (Ω_o) of these machines and the length to thickness ratio ($l/h > 10$) of the mat foundations, which renders them flexible. In practice, a foundation can be assumed as rigid if its first flexural/bending vibration frequency is 20% greater than the maximum operating frequency of the machine.

As previously mentioned, subgrade and pile impedances are computed using analytical, numerical, semi-analytical, or experimental methods. Nevertheless, independently of the method used for their computation, the following steps are used for calculating the dynamic impedance of a rigid foundation:

1. The foundation is modeled as massless and infinitely rigid; therefore, only the geometry of the area in contact with the soil is required. The use of a massless foundation is important since it avoids the need for recalculating the dynamic impedance every time that the foundation mass is modified, which often happens during the design process.
2. For each vibration mode, j, a harmonic force or torque of frequency Ω of unit magnitude is applied to the rigid foundation (e.g., $P_j(t) = P_0 e^{i\Omega t}$ or $M_j(t) = M_0 e^{i\Omega t}$), as shown in Fig. 5-9. Such force or torque generates stress waves that propagate into the underlying soil, which

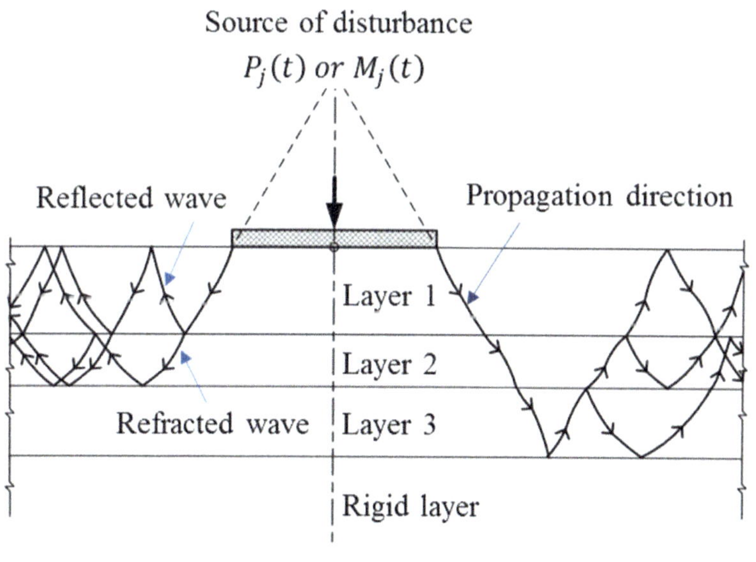

Fig. 5-9. Downward and upward wave propagation for surface disk in layered half-space
Source: Adapted from Wolf (2004).

is modeled as a viscoelastic material. Therefore, the following properties are required for each soil layer: thickness (h), modulus of elasticity (E_s) or shear wave velocity (V_s), Poisson's ratio (ν) or p-wave velocity (V_p), mass density (ρ), and material damping (β).

3. The steady state displacement response $[(U_j(t) = \bar{U}_j e^{i\Omega t - i\phi}$ *or* $\theta_j(t)] = \bar{\theta}_j e^{i\Omega t-}$ of the foundation (under the harmonic force or torque) is obtained by keeping track of the reflections and refractions that take place every time the stress waves reach a soil layer boundary. This is achieved by finding the different wave paths shown in Fig. 5-9.
4. The dynamic impedance $K_j(\Omega)$ for vibration mode j is calculated as the ratio between the harmonic force or torque acting on the foundation and its displacement response as shown in Eq. (5-26). Again, it must be noted that this is a frequency-dependent complex quantity, $i = \sqrt{-1}$.

$$K_j(\Omega) = \frac{P_j(t)}{U_j(t)} = \frac{\bar{P}_j e^{i\Omega t}}{\bar{U}_j e^{i\Omega t - i\phi}} = \frac{\bar{P}_j}{\bar{U}_j} e^{i\phi} \quad K_j(\Omega) = \frac{M_j(t)}{\theta_j(t)} = \frac{\bar{M}_j e^{i\Omega t}}{\bar{\theta}_j e^{i\Omega t - i\phi}} = \frac{\bar{M}_j}{\bar{\theta}_j} e^{i\phi} \tag{5-26}$$

5. Following soil dynamics tradition, it is customary to express the complex dynamic impedance as shown in Eq. (5-27). In addition, the real and imaginary parts of the dynamic impedance are associated, by analogy, with a dynamic (frequency dependent) *spring* and *dashpot* as shown in Eq. (5-28).

$$K_j(\Omega) = \bar{K}_j(\Omega) + i\Omega C_j(\Omega) \tag{5-27}$$

$$\bar{K}_j(\Omega) = Re\left(K_j(\Omega)\right) C_j(\Omega) = \frac{Im(K_j(\Omega))}{\Omega}. \tag{5-28}$$

6. Steps 2 through 5 are repeated for each frequency Ω of interest, until the range of vibration frequencies of the machine is covered.

In practice, the selection of an appropriate method for implementing these steps depends on the size and economics of the project, along with the availability of suitable computer codes and the level of expertise/experience in the application of computational methods for solving elastodynamics problems. In any case, the selected method should properly simulate the actual foundation geometry, underlying soil conditions, and vibration characteristics. Thus, the level of effort required to obtain a few useful impedance results may be significant, unless specialized computer codes or readily available solutions are used.

Given the extension of the topic, the discussion of computational methods and corresponding computer codes used for the calculation of dynamic impedances is outside the scope of this book; for this purpose, the reader is referred to Lysmer (1981), Wolf and Song (1996), and Wolf and Deeks (2004).

In this book, to keep the number of parameters to a minimum, the presentation is limited to simple dynamic impedance formulas and graphs covering foundations resting on or partially embedded on an elastic half-space. However, it must be noted that only deep homogeneous soil deposits can be modeled as a homogeneous half-space. Extensive compilations of foundation impedances can be found elsewhere, for instance Sieffert and Cevaer (1992) and Gazetas (1991b). However, given the variability of natural soil deposits it is very difficult to find published solutions that fit practical situations. Therefore, the main usefulness of published dynamic impedance functions, such as those discussed in Section 5.3.1 and Appendix A, is for the validation of elastodynamics computer codes, preliminary machine foundation design, and/or parametric analyses of machine foundations.

5.3.1 Dynamic Impedance Formulas for Rigid Foundations Resting on an Elastic Subgrade

Appendix A presents various formulas for calculating the dynamic impedance of rigid foundations. The information is in the form of simple algebraic expressions and dimensionless graphs pertaining to all possible (translational and rotational) modes of vibration and covering a wide range of foundation geometries on an elastic subgrade. These soil impedances are calculated and must be applied at the center of resistance (CR) of the foundation base. They can be used to determine the vibration amplitudes of block foundations under the forcing functions induced by the machine. Specific details regarding the subject formulas are provided in Appendix A, while application examples are given in Appendix C. Again, it is important to emphasize that the benchmark cases presented in Appendix A can be used for validating the computer codes used in practice for the calculation of the dynamic impedance of machine foundations. Additionally, the subject formulas can also be used for the preliminary design of flexible STG foundations.

Appendix A also introduces an approximated computational procedure, the "cone method," for the solution of elastodynamics problems. This method has a good balance of accuracy and computational effort for calculating the dynamic impedance of rigid foundations on layered soils, and unlike published solutions it is not constrained to specific soil profiles. Therefore, its range of application is only constrained by the requirement of rigid foundation behavior. For specific details of the method, including its implementation and validation, the reader is referred to Wolf and Deeks (2004).

5.3.2 Modeling of Negative Dynamic Stiffness Using Added Mass/Inertia

As discussed in previous sections, the dynamic impedance of soil- and pile-supported foundations is typically modeled using the analogue spring-dashpot system shown in Fig. 5-6(b), where the "spring" is equal to the real part of the dynamic impedance (i.e., the dynamic stiffness). However, the "spring" becomes negative for certain frequencies, as shown in Fig. 5-7 and Appendixes A and B. This results from the inertial response of the soil, as discussed in Sections 5.2 and 5.2.2. Typical structural analysis codes cannot model negative "springs" directly. Nevertheless, the negative spring sign cannot be ignored; otherwise erroneous results may be obtained. Therefore, a modified analogue/modeling approach is required in practice, to model these negative "springs." For translational DOF a mass, and for rotational DOF a mass moment of inertia are respectively added to the spring-dashpot system. For this purpose, by analogy with the SDOF system studied in Section 5.2.2, the dynamic stiffness of the subgrade for DOF j is rewritten as shown in Eq. (5-29), where $K_{j0} = Re\,(K_j\,(0))$ is the static stiffness and m_j is the added inertia used to simulate the negative dynamic "spring" for vibration frequency Ω. In general, the added inertia varies with frequency, as shown by Eq. (5-30). Therefore, multiple analyses are required to cover the range of frequencies having negative dynamic stiffness. Additional details regarding the concept of added mass/inertia and analogues to model frequency dependent impedance functions can be found in Appendix C, Section C.6, and Wolf (1994). In any case, it is important to point out that the approach discussed here is only applicable to single rigid foundations. For flexible foundations and multiple rigid foundations, it is recommended to explicitly include the soil (using solid elements) in the structural model or to use specialized SSI codes that can incorporate/model flexible foundations:

$$Re\left(K_j(\Omega)\right) = \bar{K}_j(\Omega) = K_{j0} - m_j\Omega^2 \tag{5-29}$$

$$m_j = \frac{K_{j0} - Re(K(\Omega))}{\Omega^2} = \frac{K_{j0} - \bar{K}_j(\Omega)}{\Omega^2} \tag{5-30}$$

5.4 DYNAMIC IMPEDANCE OF FLEXIBLE FOUNDATIONS

The dynamic impedance matrix is used in the dynamic analysis of machine foundations to mathematically represent the soil response. In the case of rigid foundations, this is a 6 by 6 matrix corresponding to the maximum number of DOFs required to characterize the motion of a rigid body, as shown in Section 5.3. In the case of flexible foundations, additional DOFs are required to characterize the dynamic response of the foundation and subgrade soil; therefore, significant effort is required for computing the dynamic impedance matrix of these foundations. In any case, independently of the method used for their computation (e.g., analytical, numerical, or experimental), these steps can be followed for calculating of the dynamic impedance matrix of the subgrade for flexible foundations:

1. Divide the soil-foundation system into structure and subgrade. Furthermore, identify the nodes and DOFs shared among them as shown

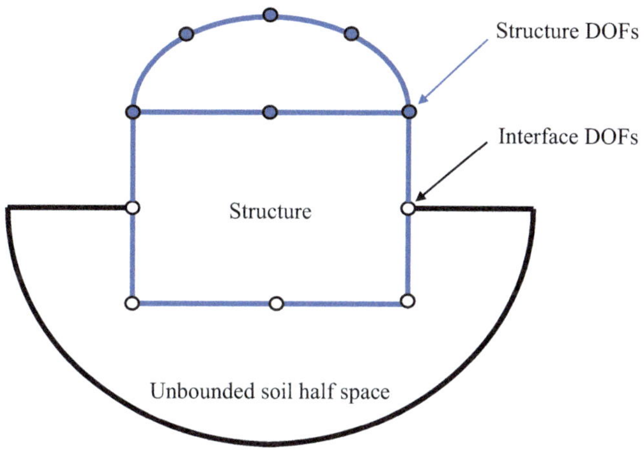

Fig. 5-10. Division of soil-foundation system substructures

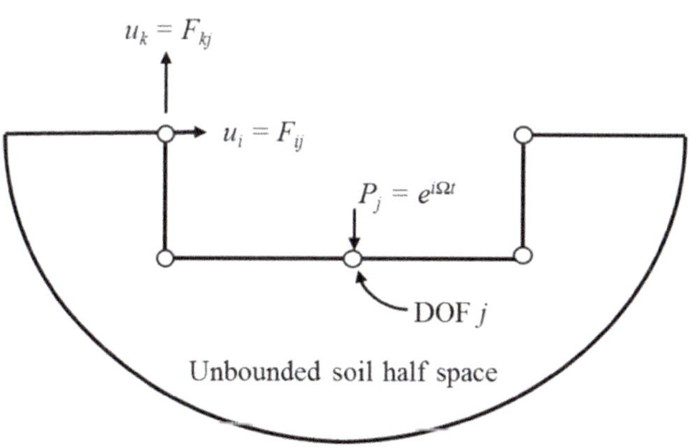

Fig. 5-11. Calculation of dynamic flexibility coefficients

in Figs. 5-10 and 5-11, where the interface nodes are shown as hollow circles.
2. Calculate the dynamic flexibility coefficients of the subgrade. As shown in Fig. 5-11, the flexibility coefficient F_{ij} is defined as the real and imaginary parts of the displacement in the direction of DOF i from a unit load of frequency Ω applied in the direction of DOF j.
3. Arrange the flexibility coefficients in matrix form as shown in Eq. (5-31), where column j is equal to the complex displacements of each interface node DOF i from a unit harmonic force acting in DOF j. This matrix is called the dynamic flexibility matrix (F).
4. Invert the dynamic flexibility matrix to obtain the dynamic impedance matrix (K) of the subgrade, Eq. (5-32).
5. Repeat steps 2 through 4 for each frequency Ω of interest, until the full range of machine frequencies is covered.

It must be emphasized that the flexibility coefficients are nothing more than the complex displacements (frequency response) of the interface DOF from a unit force of frequency Ω. Therefore, in theory, these displacements can be easily calculated using numerical (e.g., general finite or boundary element analysis codes) or analytical tools (e.g., Green's functions or integral transforms) of elastodynamics. Nevertheless, in practice, given the complexities of modeling the soil, the level of effort required to obtain a few useful impedance matrices may be significant unless specialized computer codes are used (e.g., Lysmer 1981). Furthermore, calculating the impedance matrix of the subgrade only solves half the problem, since this matrix must be added to the structural model of the foundation to calculate the complete response of the soil-foundation system. Therefore, to calculate the response of flexible foundations using elastodynamics theory, the only efficient approach is to use or develop specialized computer codes/subroutines.

$$[F(\Omega)] = \begin{pmatrix} F_{11} & \cdots & \cdots & F_{1n} \\ \cdots & \cdots & \cdots & \cdots \\ \cdots & F_{ij} & \cdots & \cdots \\ F_{n1} & \cdots & \cdots & F_{nn} \end{pmatrix} \tag{5-31}$$

$$[K(\Omega)] = [F(\Omega)]^{-1} \tag{5-32}$$

As implied here, in practice serious difficulties are faced when trying to conduct the dynamic analysis of flexible machine foundations using elastodynamics theory. One of the main issues is the lack of widely available off-the-shelf computer codes capable of solving dynamic soil structure interaction problems accounting for flexibility of the foundation mat. In the

United States, the code best known with such capabilities is SASSI (Lysmer 1981), which is widely used for the seismic analysis of nuclear power plants (NPP) but has not found its way to the design of machine foundations, other than for STGs of NPP and critical machinery in petrochemical and liquid natural gas (LNG) facilities. Therefore, when selecting the overall geometry of STG and CTG foundations, most practitioners rely on experience and rules of thumb (e.g., those provided in Chapter 3) along with the dynamic analysis of the foundation using approximated methods for modeling the subgrade. Some of these approximated methods are listed in the following, in the order of increasing degree of complexity:

1. In the case of tabletop STG foundations, a common practice is to ignore the soil response altogether and conduct a fixed base analysis of the tabletop with modal damping in the range from 2% to 5%; in other words, ignoring the beneficial effect of radiation damping.
2. Another approach used for tabletop STG foundations is to model the basemat as rigid in the structural model (e.g., using rigid constraints) and use the dynamic impedances of a rigid foundation for modeling the soil response. Additionally, in many cases, the frequency dependence of the subject impedances is ignored and the combined damping of the soil-foundation system is capped/limited in the range of 10% to 50%, depending on the engineer or company practice.
3. A modification to the previous method is to model the basemat of the machine foundation as flexible in the structural model (e.g., using shell or brick elements). Nevertheless, the dynamic impedances of the subgrade are calculated assuming that the foundation is rigid. These rigid impedances are then distributed among the translational degrees of freedom of the foundation mat based on tributary areas, the number of foundation nodes, or similar procedures (e.g., NIST 2012). Additionally, the combined damping of the soil foundation system is capped/limited in the range of 10% to 50%, depending on the engineer or company practice.

The simplifications discussed here seem to yield conservative foundation designs, when combined with previous experience. However, they are not bulletproof, given the complex behavior of flexible foundations, especially those supported on layered soils. In particular, they cannot properly capture the subgrade response of flexible foundations under vertical and rocking vibration. Therefore, the recommended approach for dynamic analysis of flexible STG and CTG foundations is to use the dynamic impedance matrix of the subgrade, considering the foundation as flexible. Nevertheless, it is recognized that simplified methods, as those described earlier, result in adequate STG and CTG foundation designs when used by experienced practitioners.

5.5 PILE-SUPPORTED FOUNDATIONS

The dynamic stiffness and damping of a pile group are affected by both the soil-pile and soil-pile cap interaction in addition to the interaction among individual piles (pile-soil-pile interaction or group effect). As a first approximation, ignoring the pile cap effect, the calculation of the stiffness for a group of identical piles may be performed in two steps. In the first step, the stiffness of the single pile is calculated considering the interaction between the pile and the surrounding soil. In the second step, the group effect is accounted for using "interaction factors" and the superposition principle.

5.5.1 Dynamic Impedance Definition for Single Piles

The dynamic impedance matrix of a single pile is calculated following procedures similar to those described for surface foundations. This matrix can be constructed using classical stiffness or flexibility methods. Here the stiffness method is used; for this purpose, consider the forces required to produce the pile head displacements shown in Fig. 5-12; the lateral force (*H*), bending moment (*M*), and axial force (*V*) from the combined effect of these displacements are given by the following expressions:

$$H = H_u u + H_\varphi \varphi \tag{5-33}$$

$$M = M_u u + M_\varphi \varphi \tag{5-34}$$

$$V = V_w w \tag{5-35}$$

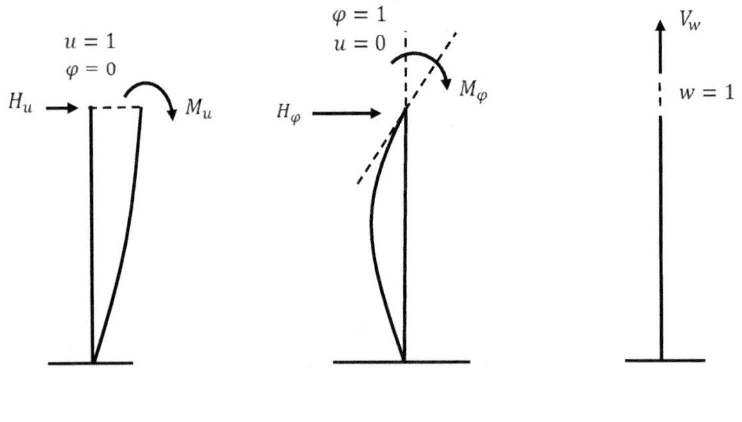

(a) Unit horizontal displacement (b) Unit rotation (c) Unit vertical displacement

Fig. 5-12. Unit pile head displacements used for calculating dynamic stiffness matrix of a single pile; (a) unit horizontal development; (b) unit rotation; (c) unit vertical displacement

The pile head impedances are directly defined from these equations. In particular, $K_{xx} = H_u$; $K_{x\varphi} = K_{\varphi x} = H_\varphi = M_u$; $K_{\varphi\varphi} = M_\varphi$, and $K_{zz} = V_w$; where K_{xx} is the horizontal force required to produce a unit displacement of the pile head while the pile head is kept fixed against rotation, and similar definitions apply to $K_{\varphi\varphi}$ and K_{zz}. Therefore, the dynamic impedance matrix for a single pile is as follows:

$$K(\Omega) = \begin{bmatrix} K_{xx} & K_{x\varphi} & 0 \\ K_{x\varphi} & K_{\varphi\varphi} & 0 \\ 0 & 0 & K_{zz} \end{bmatrix} \tag{5-36}$$

These definitions apply equally for static and dynamic loads. For the dynamic case, the forces are harmonic with frequency Ω (i.e., $P_j(t) = P_0 e^{i\Omega t}$ or $M_j(t) = M_0 e^{i\Omega t}$); the dynamic displacements are also harmonic and typically complex, given that the force and displacement in general are not in phase (i.e., $U_j(t) = \bar{U}_j e^{i\Omega t - i\phi}$ or $\theta_j(t) = \bar{\theta}_j e^{i\Omega t - i\phi}$). Using exponential notation, the coefficients of the dynamic impedance matrix are complex. The physical interpretation of these complex coefficients (real and imaginary) is as discussed for surface foundations in Sections 5.2.2 and 5.3.

Similar to the case for surface foundations, pile impedances are computed using analytical, numerical, semi-analytical, or experimental methods; the selection of the method depends on the economics of the project and the availability of appropriate computer codes. Appendix B summarizes some of the solutions available in the literature.

5.5.2 Impedance Functions of Single Piles

The pile length, bending and axial stiffness, mass, tip and head conditions, batter, and surrounding soil properties and their variation with depth and layering affect the dynamic impedance of a single pile, which in general can be described as

$$K_j(a_0) = \bar{k}_j(a_0) + i\Omega c_j(a_0) \tag{5-37}$$

where $\bar{k}_j$ and c_j are the frequency-dependent dynamic stiffness and the equivalent viscous damping coefficients for individual motions of the pile head. These coefficients are presented in Appendix B for a single pile embedded in uniform and parabolic soil profiles. The information is in the form of simple algebraic expressions, dimensionless graphs, and tables pertaining to all possible modes of vibration (translational and rotational). These pile impedances are calculated and must be applied at the pile head and can be used to determine the dynamic response of machine foundation systems for each frequency of interest.

5.5.3 Pile-Soil-Pile Interaction

When piles in a group are closely spaced, they interact with each other because the displacement of one pile contributes to the displacement of others. To obtain an accurate analysis of dynamic behavior of pile groups, it is necessary to use a suitable computer program. However, a simplified approximate analysis can be formulated based on interaction factors, a_{ij}, introduced by Poulos (1971) for static analysis and extended to the dynamic case by Kaynia and Kausel (1982). In this book, the interaction factors are defined and calculated as follows (Dobry and Gazetas 1988).

$$\alpha_{ij} = \frac{\text{Additional displacement of pile } i \text{ caused by pile } j}{\text{Displacement of pile } i \text{ under own dynamic load}} \tag{5-38}$$

To calculate the dynamic stiffness of a pile group using the interaction factors approach, the impedance functions of the single piles and their corresponding interaction factors are calculated first, then the group impedance functions are computed for each vibration frequency of the machine, as described in Appendix B and shown in Appendix C, Section C.4.

5.5.4 Pile Cap Effects

For the analysis of pile groups, it is important to determine whether the pile cap or mat provides vertical and/or lateral resistance in combination with the piles; in other words, whether contact between the soil and surface foundation elements might be expected. This will depend on the construction method and whether the long-term settlement of the soil under the pile cap is expected to be greater than the pile settlement. Loss of contact might be expected for foundations supported by end-bearing piles on soft soils, especially if surface fills are present. In such a case, no connection should be modeled between the pile cap and surrounding soil. On the other hand, for floating piles on sandy soil, pile and soil settlements can be expected to be the same. In this case, the pile cap might be considered "glued/welded" to the soil when calculating the pile group impedance. In any case, sensitivity studies considering "welded" and unconnected pile cap conditions are recommended, as described by Padrón et al. (2009).

5.6 EVALUATION OF SOIL PARAMETERS

The soil parameters required for the dynamic analysis of machine foundations include the shear modulus, G, the material damping ratio, β, Poisson's ratio, ν, and mass density, ρ. Some of the procedures that can be used to evaluate these parameters are given in the following sections.

5.6.1 Shear Modulus

The shear strains developed in the supporting soil medium caused by the dynamic loading from machine foundations are usually of a much smaller magnitude than the strains produced by static loading. The value of the soil shear modulus at smaller strains (G_{max}) is much higher than its corresponding value at larger strains. Therefore, the soil shear modulus used for the computation of the foundation impedances should be evaluated from small strain field tests (e.g., seismic cross-hole test, multi-channel spectral analysis of surface waves, MASW, seismic cone penetration test, SCPT, seismic refraction, SR, refraction of microtremors, ReMi, etc.) or from laboratory tests of undisturbed samples (e.g., resonant column test and bender element test). In the absence of measured values, there are empirical correlations with common field test measurements (e.g., standard penetration test, SPT, and cone penetration test, CPT), which can be used to preliminarily evaluate the shear modulus.

It must be noted that because of the inherent variations of in situ soils, it is necessary to complete more than one dynamic foundation analysis to bound the expected response. Therefore, the geotechnical engineer typically recommends a best estimate (BE), lower bound (LB), and upper bound (UB) for the small strain shear modulus (G_{max}). If not provided, the UB and LB values of the shear modulus can be determined from the BE shear modulus per Eq. (5-39), where COV is the coefficient of variation. As a minimum, it is recommended to take the COV as 0.3 for well investigated sites and at least 0.5 for sites that are not well investigated (i.e., lacking the required number of borings and geophysical tests per the applicable jurisdiction).

$$G_{LB} = \frac{G_{BE}}{1+COV}$$
$$G_{UB} = G_{BE}(1+COV) \tag{5-39}$$

In addition to these considerations, it must be noted that the shear modulus of the soil is also affected by the additional weight of the foundation. G_{max} increases with the mean effective confining stress. Procedures for accounting for such variation are presented by NIST (2012). Nevertheless, for typical TG foundations such effect is deemed covered by using lower bound and upper bound soil properties.

5.6.2 Material Damping Ratio

Soil material damping is a measure of energy lost from friction between soil particles during the dynamic loading. The material damping ratio can be obtained from resonant column testing, the spectral analysis of seismic waves procedure (SASW) or multi-channel analysis of surface waves (MASW). The soil material damping, at low strain, is typically 2% to 3% for sand and saturated clay.

5.6.3 Poisson's Ratio and Soil Density

The dynamic behavior of foundations is less sensitive to the values of ν and ρ. However, the dynamic impedance becomes highly frequency dependent for Poisson's ratios near 0.5. Additionally, it must be noted that the soil mass density should always be calculated from the total unit weight rather than the submerged unit weight. Total weights are used in dynamic problems because both solid and liquid phases vibrate.

REFERENCES

Dobry, R., and Gazetas, G. (1988). "Simple method for dynamic stiffness and damping of floating pile groups." *Géotechnique*, 38(4), 557–574.

Fang, H.-Y. (1991). *Foundation engineering handbook*. Chapman & Hall, London.

Gazetas, G. (1991a). "Formulas and charts for impedances of surface and embedded foundations." *J. Geotech. Engrg. Div.* 117(9), 1363–1381.

Gazetas, G. (1991b). "Foundation vibrations." *Foundation engineering handbook*, 2nd Ed., Fang, H. Y., ed. Van Nostrand Reinhold, New York.

Kaynia, A. M., and Kausel, E. (1982). "Dynamic behavior of pile groups." 2nd international conference on numerical methods in offshore piling, Austin, TX, 509–532.

Lysmer, J. (1965). *Vertical motion of rigid footings*. Waterways Experiment Station, Vicksburg, MS.

Lysmer, J. (1981). *SASSI: A system for analysis of soil-structure interaction*. University of California, Dept. of Civil Engineering, Berkeley, CA.

NIST (National Institute of Standards and Technology). (2012). *Soil-structure interaction for building structures*. GCR 12-917-21 U.S. Department of Commerce, Washington, DC.

Padrón, L. A., Mylonakis, G. E., and Beskos, D. E. (2009). "Importance of footing-soil separation on dynamic stiffness of piled embedded footings." *Int. J. Numer. Anal. Methods Geomech.*, 33(11), 1439–1448.

Poulos, H. G. (1971). "Behaviour of laterally loaded piles: II: Pile groups." *J. Soil Mech. and Found. Div.* 97 (SM5), 733–751.

Richart, F. E., Hall, J. R., and Woods, R. D. (1970). *Vibrations of soils and foundations*. Prentice-Hall, Englewood Cliffs, NJ.

Sieffert, J.-G., and Cevaer, F. (1992). *Manuel des fonctions d'impédance: fondations superficielles/handbook of impedance functions: surface foundations*. Ouest Editions, Nantes, France.

Wolf, J. P. (1994). *Foundation vibration analysis using simple physical models*. Prentice-Hall, Englewood Cliffs, NJ.

Wolf, J, P., and Deeks, A. J. (2004). *Foundation vibration analysis: A strength-of-materials approach*. Elsevier, Amsterdam.

Wolf, J. P., and Song, C. (1996). *Finite-element modelling of unbounded media*. Wiley, Chichester, UK.

CHAPTER 6
FINITE ELEMENT MODELING

6.1 INTRODUCTION

Finite element analysis (FEA) methods are numerical procedures for solving solid and continuum mechanics problems. Advanced and powerful commercial structural analysis programs using finite element method (FEM) are widely used in the modeling, analysis, and design of various types of structures, including TG foundations. This chapter provides general guidelines for developing FE models for TG foundations. Given the unique geometries and structural characteristics of TG foundations, emphasis is placed on selection of appropriate element type, mesh size, modeling of machine mass, and soil or pile boundary conditions. Topics on soil-structure interaction effects are provided in Chapter 5 and Appendixes A through C.

It is essential to understand that the quality (accuracy) of an analysis depends largely on the user's knowledge on FEA theory and modeling features available in the computer program chosen. It should be noted that this chapter does not cover fundamental FEA and basic modeling features in various computer programs.

TG foundations are typically analyzed by using 3D FE models rather than 2D models because of their complex configurations. It should be noted that this chapter focuses on 3D FE models, but the guidelines apply to 2D models as well.

6.2 GENERAL MODELING CONSIDERATIONS

The first step for developing a structural model is to create a finite element model representing appropriate structural geometry based on the foundation's preliminary layout described in Chapter 3. This structural model contains a series of nodes (points), members (lines), and various types of elements (i.e., beam, shell, and solid elements) to represent the geometry of the actual structure. In practice, it is useful to create a geometry model based on the centerline of the structural members and/or elements.

When analyzing a TG foundation system, several key components should be considered as listed here (see Fig. 6-1 for an example).

- Vibrating machinery;
- Machine-foundation interface;
- Foundation-ground media interface;
- Soil/ground medium.

The following generic guidelines should be considered during the development of an FE model for the analysis and design of a TG concrete foundation structure:

1. The finite element mesh should be dominated by quadrilateral and hexahedral elements with aspect ratios less than or equal to 2 and without edge, angular, or parallel distortions or warping.
2. The constant strain plane triangular element and the constant strain solid tetrahedral element are not recommended. However, the six-node plane triangular element and the ten-node solid tetrahedral element, shown in Fig. 6-2, are accurate and numerically efficient.

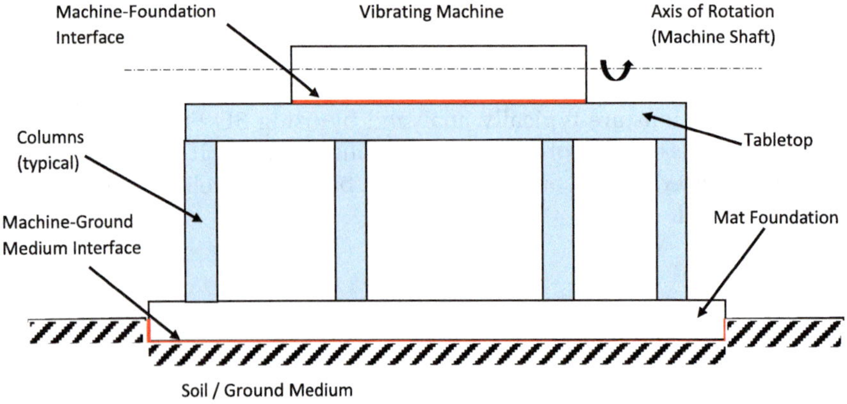

Fig. 6-1. An elevated space-frame pedestal foundation

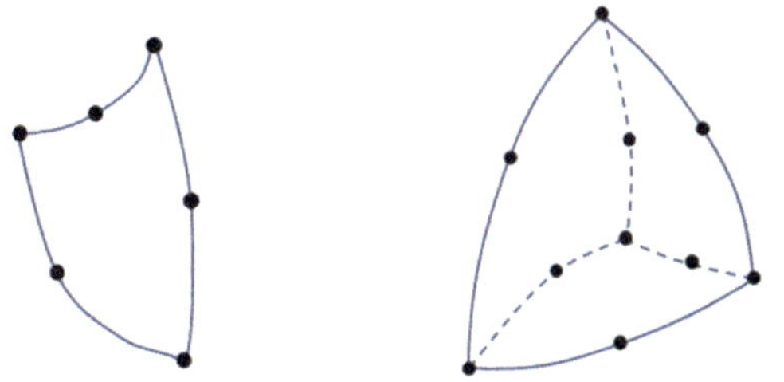

Fig. 6-2. High order triangular and tetrahedral elements

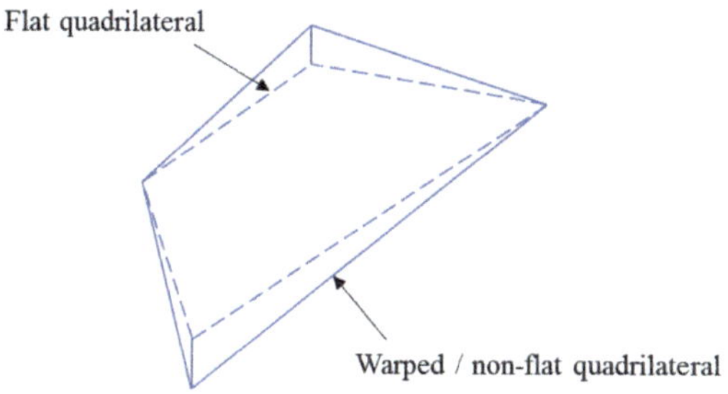

Fig. 6-3. Sample schematic presentation of warping condition in a 2D element

3. The interior corner angles of a plate/shell/solid finite element (2D or 3D) should not be less than 30 degrees or greater than 150 degrees to minimize distortion effects.
4. The maximum warping angle, defined as an angle between the two normal vectors that are perpendicular to the planes that pass through two different three-node sets of a quadrilateral element, should be less than 10 degrees. Some 2D plate/shell element formulations are based on flat planar geometry and they should lie in a flat surface without warping (see Fig. 6-3).
5. Stress singularities and concentrations exist at the following locations:

   - corners of openings;
   - any sudden geometrical changes in the FE mesh;
   - locations of concentrated loads; and
   - areas close to supports and boundary conditions.

   A very fine mesh solution would indicate very large stresses, and further mesh refinement would produce higher stresses. However, as a result of concrete cracking, these high stresses are actually much lower. Hence, the FE model should predict the overall behavior of the concrete structure, while stress concentrations can be addressed at the design level by providing diagonal reinforcement at opening corners, or by modeling the concentrated loads uniformly over the area of application.

6. Shear deformation needs to be considered for thick walls, slabs, and mat foundations used in TG foundations. Therefore, as a minimum, shell elements based on first-order shear deformation theory (i.e., thick plates) should be used for such structures.
7. Some FE programs report element outputs only at the integration Gaussian points and/or the element centroid. Therefore, the mesh should be tuned for reporting results at desired critical design locations (e.g., support faces, middle of the span, at "d" distance from the support, where "d" is a member's effective depth). Alternatively, stress contours and interpolation/extrapolation may be used to evaluate stress resultants at desired critical locations. For detailed information on the transformation of output for solid elements, refer to Foster et al. (2003) or Hoogenboom and de Boer (2010).
8. The design engineer should understand the general characteristics, applicability, and limitations of finite elements used in modeling TG foundations, and pay special attention when connecting different types of elements, such as at transition zones between beam elements, shell elements, and solid elements, to ensure a consistent and compatible FE model. Along any common edge or face of the adjacent elements, there should be one-to-one correspondence between common nodes and their corresponding degrees of freedom. Ensuring proper compatibilities between common nodes, edges, and faces of adjacent elements helps create an integral FE model.

Detailed information on various formulations and modeling of finite element method can also be found in Wilson (1998), Bathe (1996), and Zienkiewicz and Taylor (2000a, b).

6.3 ELEMENT SELECTION

The selection of the element type is an important decision the design engineer needs to make. Selecting a particular element type implicitly means accepting the element interpolation functions and the assumptions used to derive the differential equations and element matrices. TG foundations are usually modeled using beam, shell, and solid elements, depending on their geometric characteristics and the purpose of an analysis.

6.3.1 Beam Elements

The beam element is a two-node, one-dimensional line element, which is the most common element used in the analysis of frame structures. It can be used for modeling pedestals, beams, girders, and columns. In a 3D FE model, a beam element has six degrees of freedom (DOF) at each node (three translational DOFs and three rotational DOFs). The following sectional properties are required as input data for a computer model:

- Cross-sectional area;
- Moments of inertia about two principal local axes of the cross section;
- Shear areas; and
- Polar (torsional) moment of inertia.

Elevated TG tabletops have deep sections and short spans where shear deformation cannot be ignored. If beam elements are chosen to model tabletop members, the design engineer should specify the shear areas to include the shear deformation effects. Otherwise, the results, particularly the deflections, may not be accurate since the shear deformation is ignored.

Rotational and/or translational degrees of freedom at the end nodes of a beam element can be released to simulate appropriate compatibilities with adjacent elements as required.

6.3.2 Two-Dimensional Plate/Shell Elements

There are several theories for the analysis of deflection and stresses of shell elements. Plate theories are traditionally divided into "thin" (Kirchhoff-Love) and "thick" (first-order shear deformation theory or Mindlin-Reissner) plate theories. For TG foundations, thick plate formulations are recommended, since shear deformation needs to be considered. Shear deformation tends to be important when shell thickness is greater than approximately 1/10 of the span of plate-bending curvature.

Plate/shell elements are generally defined as quadrilateral (4 nodes or 8 nodes) or triangular (3 nodes or 6 nodes) two-dimensional elements with five to six degrees of freedom per node. These elements are commonly used for modeling walls and mat foundations. They can represent out-of-plane bending and in-plane membrane behaviors. However, in-plane rotational (or drilling) stiffness is typically not considered or only approximately modeled. Some programs automatically add a small in-plane rotational stiffness; others use variational principles to approximate such stiffness. The design engineer should understand such limitations when using shell elements or connecting shell elements to the other types of elements.

6.3.3 Three-Dimensional Solid Elements

Three-dimensional (3D) solid elements are the most generic type of finite elements used to model a linear/nonlinear continuum.

They can be used to account for shear deformation effects and nonuniform distribution of thermal gradients across thick walls. The required number of layers of solid elements across the thickness of structural components depends on the order and compatibility of the elements and loading types.

The basic outputs of these elements are stresses and displacements at the integration points of the elements. Many commercial FE programs also calculate resultant forces at element centroids, nodes, or along a cut through multiple elements. The design engineer can investigate the distributions of principal stresses to understand or verify the load transfers between different structural components. The sectional forces and moments required for reinforcement design can be computed by performing stress integration across the member thickness. For detailed information on the transformation of output for solid elements, refer to Foster et al. (2003) or Hoogenboom and de Boer (2010).

6.3.4 Spring/Damper Elements

Spring elements are 2-node elements with spring constants or coefficients in the desired directions for static and dynamic analyses. Similarly, damper elements (for dynamic analysis) are 2-node elements with damping constants or coefficients in the desired directions. Thus, for dynamic analyses, combined spring-damper elements or independent spring or damper elements can be used for simulating the stiffness and damping of foundation boundary interfaces, such as soil-supported, pile-supported, or vibration isolated (Chapter 10) TG foundations. Usually, these elements are attached to nodal points of the mat foundation. Spring element output, in the form of internal forces, moments, and displacements, can be used for calculating soil pressures and foundation settlements. In some FE programs, spring or damping characteristics may be modeled directly at the nodal boundary without additional spring elements.

TG foundations have large structural components such as pedestals, columns, and tabletops supported on very thick mat foundations. Therefore, the spring elements under the mat foundation usually do not experience uplift conditions. In practice, there is no need to perform nonlinear analysis for such foundations using compression-only springs.

6.3.5 Rigid Link Elements

When beams, columns, walls, and mat foundations are modeled using beam and shell elements, the joint intersection sizes should be modeled using rigid links. This is necessary to account for the effect of "short span and large joint size" (Fig. 6-4).

Rigid links can be modeled using beam elements with a much larger Young's modulus than other elements in the model (e.g., 100 times stiffer than the elements they link). Rigid links can also be used in transferring loads at machine shaft level to tabletop, or from tabletop surface to beam element centerline (Fig. 6-5).

If 3D solid elements are used to model a tabletop, it may be necessary to define rigid surfaces as well (i.e., constrain equation or rigid surface feature in some FE programs) for the areas covered by soleplates, in order to capture the restraint effects induced by the machine anchorage. Fig. 6-6 is an example of such rigid surfaces. This modeling technique can be used to avoid singularity-type deflections at just a few individual nodes where loads are applied.

When using a computer program to automatically calculate the structure masses from element properties, rigid link material properties must be carefully selected to reflect the actual structure mass distributions. For

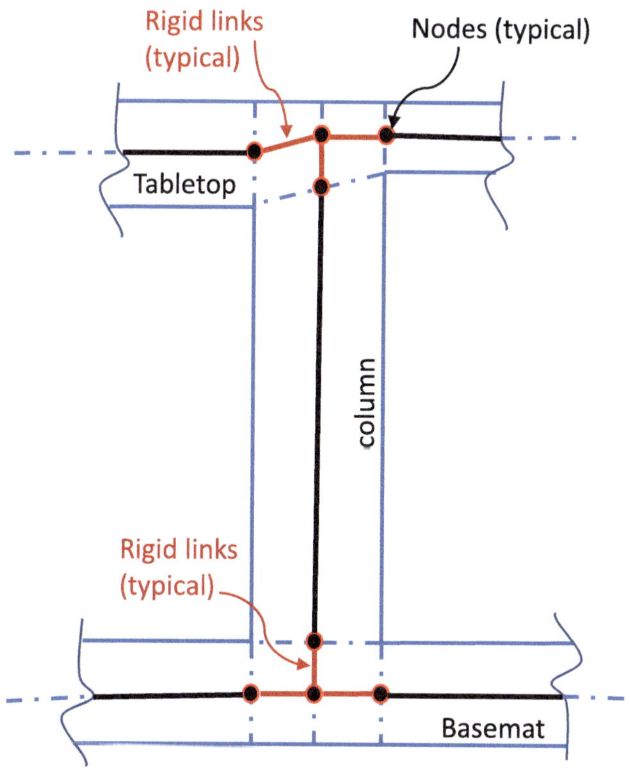

Fig. 6-4. Example of rigid links (beam elements) in a TG FE model

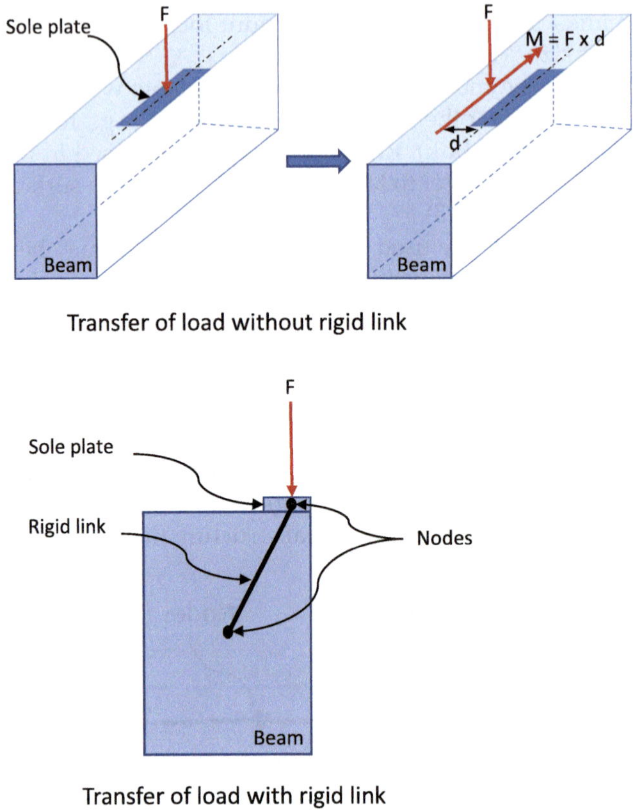

Fig. 6-5. Beam element modeled with and without rigid links

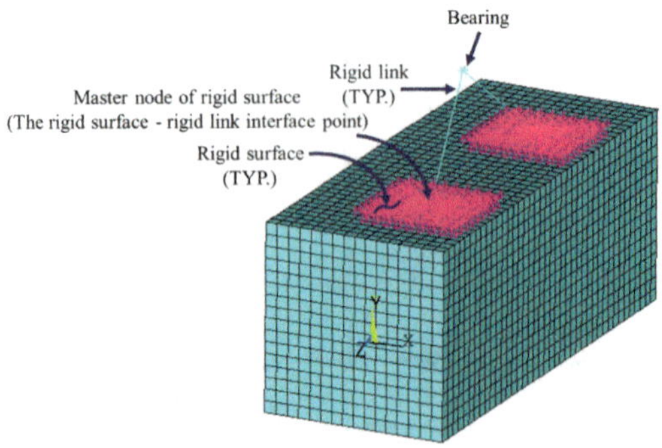

Fig. 6-6. Rigid surfaces to model soleplates

example, at member intersections, multiple-counting of masses from each interconnected member should be avoided.

6.3.6 Element Selection Recommendations for TG Foundations

For elevated space-frame pedestal foundations, it is recommended to use beam and/or shell elements to model the whole structure for strength design purposes, since it is efficient to design reinforcement based on approximate, conservative member forces and moments. Fig. 6-7 is an example of this type of analytical model. However, for dynamic and static deflection performance analyses, it is recommended to use 3D solid elements to model the entire structure, or at least the tabletop portion, while columns, walls, and mat foundation can be modeled using beam and shell elements. Fig. 6-8 is an example of this type of analytical model.

For TG block foundations, such as gas turbine foundations or side-exhaust or axial-exhaust steam turbine foundations, 3D solid elements should be used to model the entire structure for the dynamic and static deflection performance analyses. Fig. 6-9 is an example of this type of FE model. For reinforcement design, using a combination of different element types (beam, shell, 3D solid elements) is sufficient.

6.4 MESH SENSITIVITY

Finite element analysis results are known to be sensitive to the element type (e.g., shell or solid elements), element shape (e.g., quadrilateral, triangular, prism, pyramid, tetrahedral, or hexahedral), and mesh refinement. Therefore, mesh sensitivity studies are necessary unless a proper mesh size is already known in order to use the FE model for the actual structural design. For example, Kim and Moreschi (2013) performed a sensitivity study and determined the required finite element mesh sizes to generate proper dynamic responses of a concrete structure.

The following general recommendations should be considered for selecting proper mesh size for FE models of TG pedestal foundations:

- Minimum three layers of "first-order elements" should be used across the thickness of any structural components.
- For transverse bending in beams, maximum element size should range from "span length/6" to "span length/8."
- For transverse bending in columns or walls, a minimum of four elements should be used across the clear span.
- If concentrated loads are important, arrange more elements around the concentrated loads to properly calculate the punching shear at the critical sections.

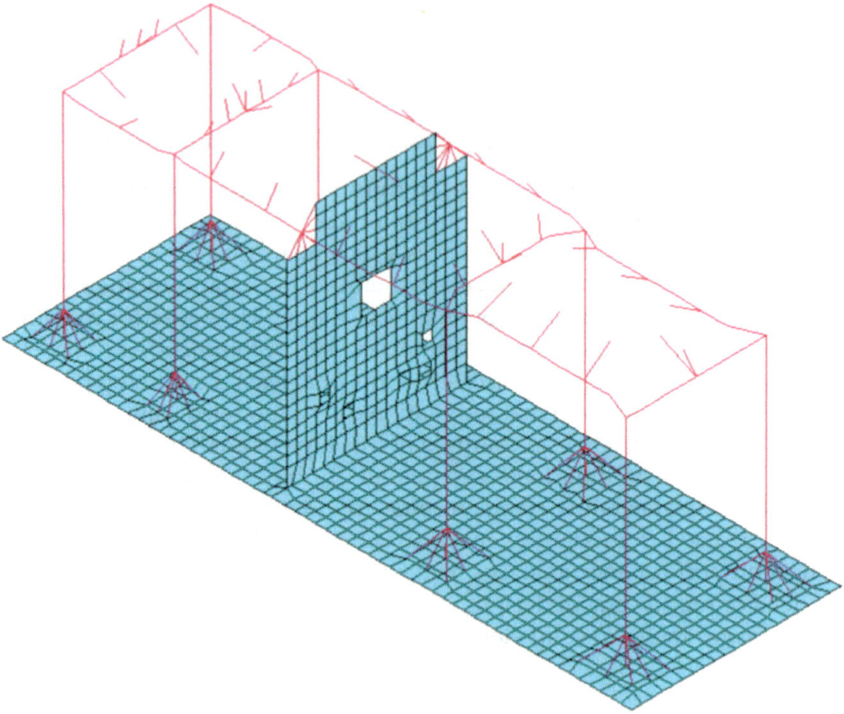

Fig. 6-7. Elevated space-frame STG pedestal foundation FE model using beam and shell elements (with rigid links)

The sensitivity analysis and evaluation in two consecutive refined meshes can provide sufficient verification for the adequacy of an FE mesh.

6.5 GEOMETRY MODELING

TG foundations usually have complex geometries. Small geometric details such as notches, pockets, and cutouts need not be modeled. However, large openings, sizable trenches, and masses that can significantly affect the structural stiffness and mass distribution of the system need to be included in the model.

The dynamic model of a TG foundation should include the effect of the TG machine set. Omission of the TG machine set would introduce considerable error in the calculated response of the TG foundation. There are two acceptable methods for modeling the TG machine set.

The first method is to use lumped masses at designed soleplate locations, based on dead weight at each soleplate location provided by the

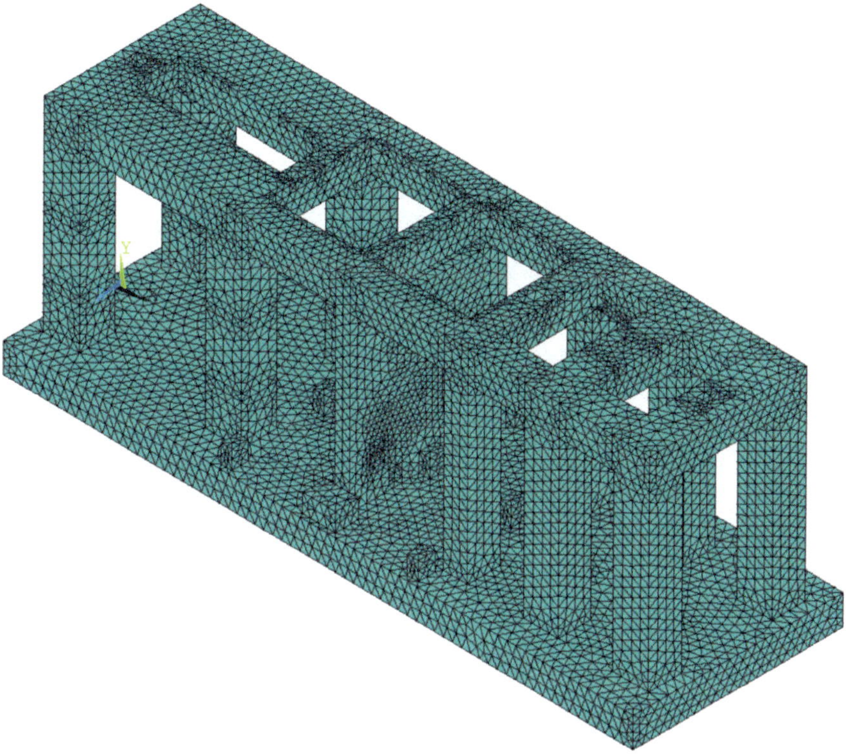

Fig. 6-8. Elevated space-frame STG pedestal foundation FE model using 3D solid elements

machine manufacturer. Another alternative is to lump machine masses at the shaft elevation and connect them to support locations at top of concrete using rigid links. Though the dynamic interaction between the TG machine set and the foundation is not taken into account, this method is commonly used in practice and acceptable to many TG manufacturers.

The second method is to explicitly model both the TG machine set and the foundation. This involves modeling the shaft as a lumped mass-beam system, modeling the bearing oil film as spring elements with specified stiffness and damping characteristics, and modeling the bearing pedestals as spring-mass systems, all of which are connected to the FE model of the foundation. If TG manufacturers provide the rotor section and mass distribution details to foundation designers, the rotor train can be modeled as beam elements based on the provided details. The stiffness and mass distribution of the turbine shell should also be incorporated if information is available from the manufacturers. Though a more accurate representation of a TG machine-foundation system, this method is seldom implemented

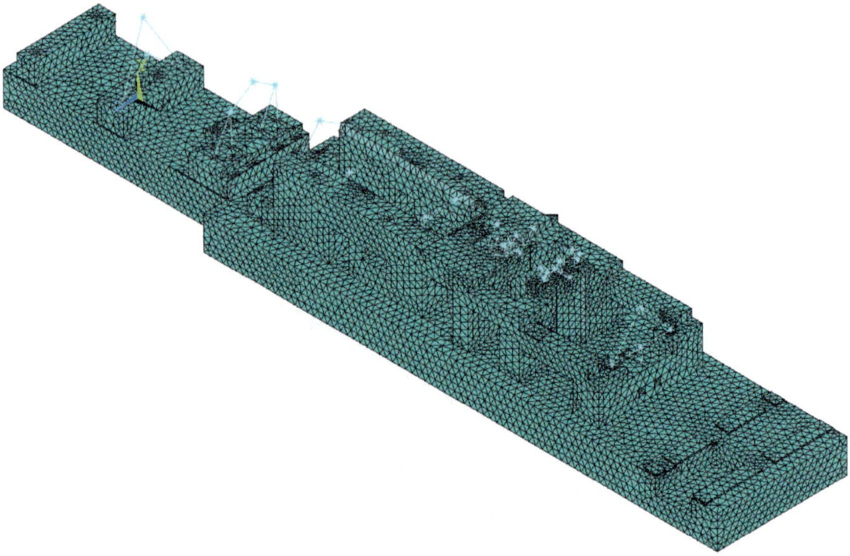

Fig. 6-9. Single shaft STG-CTG block foundation FE model using 3D solid elements

in practice, because detailed machine design data (stiffness, damping, mass) is usually not available to the foundation design engineer, and even if such data are available, the extra effort required to do so can make it economically unfeasible.

6.6 BOUNDARY CONDITION MODELING

For static analysis purposes, interactions between the mat foundation, whether treated as rigid or flexible, and the supporting subgrade or piles can be modeled using a system of vertical and horizontal spring elements (boundary spring elements). For dynamic analysis purposes, a set of dynamic impedances (spring-damper systems) can be used to model the subgrade response for rigid foundations (see Section 5.3), whereas a dynamic impedance matrix is required for flexible foundations (see Section 5.4).

6.6.1 Boundary Condition Modeling for Static Analysis

For static analyses of TG foundations, such as overall differential settlement calculations or strength designs, stiffnesses of spring elements can be calculated using traditional subgrade reaction or static pile stiffness provided by the geotechnical engineer, as they are frequency-independent.

However, for various differential deflection/rotation criteria checks (such as MTM. See Section 7.4.2) during normal operational loading conditions, for which the permanent dead loads are usually excluded, the soil or piles will probably respond in a much stiffer manner. This is because the induced soil strains under such loads are usually very low (less than 0.005%). Therefore, the deformation is almost linear elastic and can be approximately simulated using subgrade modulus based on the "tangent-at-the-origin" shear modulus, or "dynamic" shear modulus. Therefore, the foundation designer should consult a geotechnical engineer for appropriate stiffness to be used for these evaluations.

For soil-supported flexible mat foundations, the spring constant at each node can be obtained by dividing total stiffnesses (vertical and horizontal) by total number of nodes, or using the contributory area of each node (Fig. 6-10). Note that the side soil around the mat foundation, that is, the foundation embedment effect, is usually ignored.

Besides the translational (vertical and horizontal) stiffnesses, the soil also exhibits global rocking (i.e., rotation about the longitudinal and transverse axes) and torsional (i.e., rotation about the vertical axis) behaviors. Since the translational springs already contribute to the rocking and torsional stiffness, three rotational stiffness deficits (two rocking, one twisting) can be calculated as required.

To capture the effect of this stiffness deficit, the FE model may be supplemented with rotational springs. The total rocking deficit about each horizontal axis can be evenly distributed along that axis, while the total torsional deficit about the vertical axis can be applied at the geometric center of the mat foundation. There are other alternative approaches to capture the stiffness deficit effects, for example, using stiffer spring elements along boundaries of the mat foundation. Refer to Harden et al. (2005) for further discussions on these techniques.

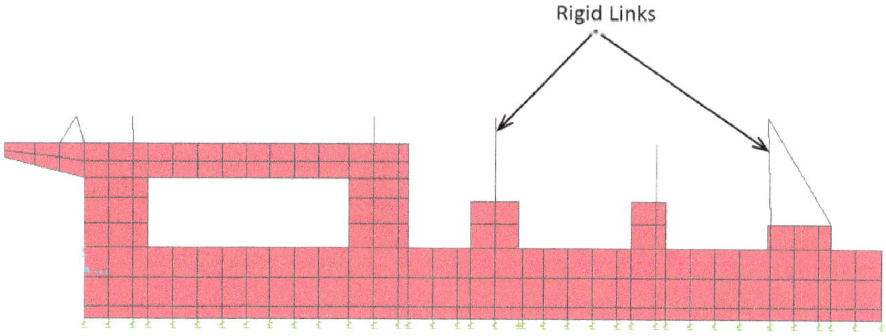

Fig. 6-10. Soil springs underneath a TG foundation

For pile-supported mat foundations, the common practice for static analysis and design is to use three spring elements (two horizontal and one vertical) at each pile location. The moment restraints at the pile tip, whether fixed-head or free-head, can be ignored.

6.6.2 Boundary Condition Modeling for Dynamic Analysis

For dynamic analyses of TG foundations, determining the proper dynamic impedances, including damping, of subgrade or pile can be very challenging, since such impedances are frequency-dependent in nature. Chapter 5 discusses how dynamic impedances and damping can be calculated and implemented in FE analyses. This section only briefly discusses common practices in how to model the dynamic impedances in an FE model. See Sections 5.3, 5.4, and 5.5 for more detailed discussions on, for example, modeling of soil damping and negative dynamic impedance.

For soil-supported rigid block mat foundations, the calculated dynamic impedance (see Chapter 5) can be modeled using six sets of stiffness and damping properties, three translational and three rotational, at the geometric center of bottom of the basemat (Fig. 6-11). In addition, two coupled stiffness and damping terms (off-diagonal terms in the stiffness/damping matrix) can be specified to account for coupling between global mode shapes.

For soil-supported flexible mat foundations, different approaches are used in practice for modeling the soil response, as discussed in Section 5.4. However, in general, the recommended approach is to use a substructuring

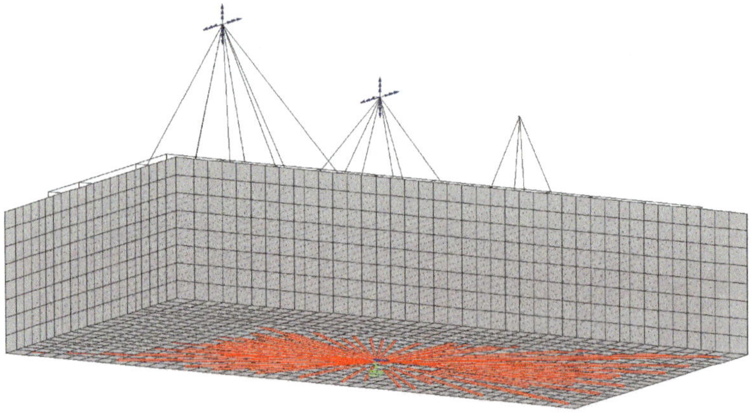

Fig. 6-11. Soil springs at bottom geometric center of a rigid TG foundation Source: Courtesy of WorleyParsons; reproduced with permission.

method, where the soil response is captured using the dynamic impedance matrix of subgrade. Alternatively, the soil can be included explicitly in the structural model, for instance, using solid elements. However, the computational cost of such an approach is usually high, given that the domain of the soil mesh must be sufficiently large to avoid boundary effects, unless special purpose elements are used at the soil mesh boundaries to prevent wave reflections. Specific details of this explicit modeling approach can be found in Wilson (1998) and ASCE 4.

The pile or pile group dynamic impedances should be calculated per Chapter 5 recommendations.

6.6.3 Boundary Condition Uncertainty

A range of pile or soil stiffness properties, away from nominal values, should be investigated to account for the uncertainty or variability of such design parameters. The geotechnical engineer should provide reasonable upper and lower bound stiffness values to the foundation design engineer. In the absence of any geotechnical guidelines, see recommendations in Chapter 5. For example, Eq. (5-39) recommends ± 30% for well-investigated sites and ± 50% otherwise.

6.7 MASS MODELING

For dynamic analysis, all major masses of the machine-foundation system must be accounted for in the FE model.

The mass of each structural element of the TG foundation is calculated and lumped at the nodal points by the FE computer program. Element dimensions and material properties should, therefore, be selected to reflect the actual mass distributions of the structure. Some computer programs have features that can either add additional masses or directly define masses by using mass elements. The design engineer should check the computed mass output against manual calculations to ensure that all masses are appropriately captured for the dynamic analysis. The mass of rotating machine parts (rotors) can be lumped to nodes distributed along the shaft center line at each bearing location. If the tabletop is modeled using 3D solid elements, the masses of machine stationary parts can be accounted for by adjusting the mass density of the volumes that represent soleplates using the equation that follows:

$$\rho_i = \rho_c + \frac{W_{mi}}{V_i} \tag{6-1}$$

where ρ_i is the density assigned to the contact volume, ρ_c is the density of concrete, and W_{mi} is the machine mass acting on the contact volume V_i.

6.8 LOAD MODELING

Forces and moments are applied as nodal loads or member loads, or distributed to nodes within a volume or on a surface. In cases where forces are eccentric to the nodal point, a combination of moments and forces can be utilized for proper load transfer.

Dynamic rotor loads can be applied to nodal points along the shaft centerline at each bearing location, which is connected to the tabletop using rigid links. Alternatively, they can be applied to the soleplates as equivalent force couples, considering the height differences between shaft centerline and top of the machine soleplates.

6.9 DAMPING MODELING

Damping reduces response motions of a structural system subjected to dynamic loads. It reflects the energy dissipation mechanisms of the system. Damping ratio is expressed as the percentage of damping relative to critical damping. The critical damping is the amount of damping that will produce no system oscillation.

In actual structures, many mechanisms contribute to the dissipation of energy. In practice, the damping in actual structures is usually represented in a highly idealized manner, such as a linear viscous damper or dashpot. The damping coefficient is selected so that the vibrational energy it dissipates is equal to the energy dissipated in all damping mechanisms combined, presented in the actual structure. This idealization is called equivalent viscous damping (Chopra 1995).

For steady-state harmonic response of a system, such as a TG foundation under operational dynamic loads, it is recommended to specify 2% of critical damping for each structural mode considered in a dynamic analysis (i.e., structural modal damping). This lower than normal damping value reflects the fact that concrete TG foundations are, in general, stressed to rather low levels under operating cyclic loads. Higher damping values, such as 7% for reinforced concrete TG foundations, could be used for scenarios like emergency loading to reflect the fact that the foundation may be stressed to higher levels than under normal operating conditions, if dynamic analysis is required for such conditions.

Inclusion of soil and pile damping effects, if deemed necessary, in an FE model for a TG foundation, can be very complicated. See Chapter 5 for details.

6.10 SECTIONAL AND MATERIAL PROPERTIES

Section properties should reflect linear elastic behavior for both static and dynamic loads. Large TG concrete foundations should use minimum 4,000 psi strength concrete. With improved concrete quality over the years, and because the TG foundations are governed by stiffness requirements, the stress levels in a TG foundation are usually very low in day-to-day operations, which results in minimal concrete cracking. Therefore, it is satisfactory to assume uncracked concrete property elements for both dynamic and static analyses.

The modulus of elasticity of concrete depends on the time and rate of loading. The static and dynamic modulus of elasticity can be calculated according to applicable design codes and standards (e.g., ACI 318, ASTM C469, and ASTM C215). The concrete dynamic modulus of elasticity is always greater than the static modulus of elasticity, and it is based on the composite nature of concrete, rather than nonlinear behavior of concrete exposed by varying strain levels. It is recommended that an increase be considered in Young's modulus (in the range of 10% to 60%) for dynamic analyses only, per ACI CRC #43 (Popovics 2008), as well as ACI 351.3R.

6.11 MODEL VERIFICATION

Once the FE model is constructed, it is necessary to perform various verifications to confirm that the model is appropriate for its intended purposes. Following are some general guidelines when performing such model verifications:

1. Check for unconnected nodes (i.e., no "floating" nodes).
2. Check for element connectivity (i.e., no multiple structures).
3. Verify that dimensional and force/moment units are consistent.
4. Verify that the total structure weight modeled matches the actual weight calculated by hand.
5. Review equilibrium and displacements around the applied concentrated loads, supports, and boundary conditions.
6. Perform simple static analyses with unit loads at selected nodes/elements to check the overall deformed shape, and verify that the deflections, soil settlements, and reactions match hand calculations.
7. Refine the model using finer meshes and recheck the results of interest to ensure the mesh sizes are appropriate.

6.12 AVERAGING OF FE RESULTS

Appropriate averaging of FE results for foundation design is a matter of engineering judgment. It is the responsibility of the foundation design engineer to determine what methods will be appropriate. Technical manuals of FE software usually provide some information on this topic. For example, one important design consideration is to determine the design strip width for mat foundation design. If the design strip is too large, the moments and shears will be averaged over too wide a region, causing an unconservative design. On the other hand, if the strip width is too small, the moments and shears will be overestimated and the design will be overconservative. Generally, there are three methods for determining the design strip width:

1. *ACI Strip Method*—Per Section 8.4.1 of ACI 318, a column strip is a design strip with a width on each side of a column centerline equal to 25% of the span length or width, whichever is smaller. A middle strip is a design strip bounded by two column strips.
2. *Zero-Shear-Transfer Method*—This method is explained in PTI DC20.6-99. It uses the shear contours to identify where the shear changes sign. Then the design strip width is determined based on the distance between points of zero shears. This method should provide a design strip similar to that using the ACI strip method. However, since it is derived from a theoretical solution, it is more suitable for uneven column layouts.
3. *Shear Perimeter Method*—The shear perimeter method is discussed in NIST GCR 12-917-22. It considers a design width equal to the pedestal width plus a distance of half to one mat thickness on each side. This will lead to a more conservative design for flexure than the preceding two methods. This method is more applicable for cases where shear or punching shear failures are a primary concern.

Besides these three methods, a simple rule of thumb practice is to use a strip width no more than 3 times the cross-section depth. This usually gives a satisfactory design unless further refinement to reduce reinforcement is required.

REFERENCES

ACI (American Concrete Institute). (2014). "Building code requirements for reinforced concrete." *ACI 318-14*, Detroit.

ACI. (2004. Reapproved 2011). "Foundations for dynamic equipment." *ACI 351.3R-04*, Detroit.

ASCE. (2016). "Seismic analysis of safety-related nuclear structures." *ASCE 4-16*, Reston, VA.

ASTM International (American Society for Testing and Materials). (2008). "Standard test method for fundamental transverse, longitudinal, and torsional resonant frequencies of concrete specimens." *ASTM C215*, West Conshohocken, PA.

ASTM. (2002). "Standard test method for static modulus of elasticity and Poisson's ratio of concrete in compression." *ASTM C469/C469M*, West Conshohocken, PA.

Bathe, K. J. (1996). *Finite element procedures*. Prentice-Hall, Upper Saddle River, NJ.

Chopra, A. K. (1995). *Dynamics of structures: theory and applications to earthquake engineering*. Prentice Hall, Upper Saddle River, NJ.

Foster, S. J., Marti, P., and Mojsilovic, N. (2003). "Design of reinforced concrete solids using stress analysis." *ACI Struct. J.*, 100(6), 758.

Harden, C., Hutchinson, T., Martin, G. R., and Kutter, B. L. (Aug 2005). *Numerical modeling of the nonlinear cyclic response of shallow foundations*. PEER Report 2005/04, Pacific Earthquake Engineering Research Center, College of Engineering, University of California, Berkeley, CA.

Hoogenboom, P. C. J., and de Boer, A. (2010). "Computation of optimal concrete reinforcement in three dimensions." *Computational modelling of concrete structures*. Taylor & Francis, London.

Kim, H. K., and Moreschi, L. (Aug 2013). "Finite element model of interior concrete structures for seismic analysis with high-frequency content." *Transactions of the 22nd International Conference on Structural Mechanics in Reactor Technology (SMiRT-22)*, San Francisco, CA.

NIST (National Institute of Standards and Technology). (2012). "NEHRP seismic design technical brief no. 7: seismic design of reinforced concrete mat foundations." *NIST GCR 12-917-22*, Gaithersburg, MD.

Popovics, J. S. (Concrete Research Council). (2008). *CRC 43: a study of static and dynamic modulus of elasticity of concrete*. Farmington Hills, MI.

PTI (Post-Tensioning Institute). (1999). *Design fundamentals of post-tensioned concrete floors. PTI DC20.6-99*, Farmington Hills, MI.

Wilson, E. L. (1998). *Three-dimensional static and dynamic analysis of structures: a physical approach with emphasis on earthquake engineering*, 2nd Ed., Computers and Structures, Walnut Creek, CA.

Wilson, E. L., Taylor, R. L., Doherty, W., and Ghaboussi, J. (1971). "Incompatible displacement models," Proceedings, ONR Symposium on Numerical and Computer Methods in Structural Mechanics, University of Illinois, Urbana.

Zienkiewicz, O. C., and Taylor, R. L. (2000a). *The finite element method, Vol. 1: the basis*, 5th Ed. Butterworth-Heinemann, Oxford, UK.

Zienkiewicz, O. C., and Taylor, R. L. (2000b). *The finite element method, Vol. 2: solid mechanics*, 5th Ed. Butterworth-Heinemann, Oxford, UK.

CHAPTER 7

SERVICEABILITY ANALYSIS AND ACCEPTANCE CRITERIA

7.1 INTRODUCTION

This chapter covers the dynamic and static serviceability analyses and acceptance criteria required by turbine generator manufacturers for the TG foundation design. TG foundations are subjected to various loads, some static, some resulting from dynamic conditions, others resulting from shrinkage and thermal expansion and contraction as outlined in Chapter 4. The purpose of the serviceability analyses is to determine whether a proposed foundation will meet the serviceability acceptance criteria imposed by the project design specifications under the application of these loads.

Two types of serviceability criteria shall be satisfied:

1. Dynamic serviceability criteria shall be met to ensure that the foundation system will not experience resonance or excessively high vibration amplitudes during normal operations.
2. Static serviceability criteria shall be met to ensure that the TG bearing alignment will be maintained during operation.

It should be noted that serviceability acceptance criteria vary from manufacturer to manufacturer but are in general stringent, causing TG foundations to be larger than required by the strength design criteria imposed by the applicable building codes.

7.2 DYNAMIC SERVICEABILITY ANALYSIS

A typical procedure for the dynamic serviceability analysis includes the following:

1. Collect dynamic design input and acceptance criteria.
2. Determine dynamic unbalance loads and load combinations.
3. Calculate dynamic soil or pile impedances.
4. Create dynamic FE model of TG foundation system.
5. Perform dynamic serviceability analysis.

7.2.1 Collect Dynamic Design Input and Acceptance Criteria

Three types of design input should be collected for the dynamic serviceability analysis.

Geotechnical Design Information. The geotechnical data are used to evaluate the soil or pile stiffness and damping coefficients. The soil or pile parameters listed in Section 3.2.2 should be collected from geotechnical consultants for the dynamic serviceability analysis.

TG Design Information. The TG machine information listed in Section 3.2.2 should be collected from the machine manufacturer for dynamic serviceability analysis. The dynamic loads and load combinations of TG machines will be discussed in the next section.

Different machine manufacturers may have different dynamic design requirements for the TG foundation, but these requirements typically include an acceptable foundation vibration limit in terms of amplitudes, velocity, or accelerations. Some manufacturers provide the TG machine's damped first and second natural critical speeds, and require that the local natural vibrating frequencies of the structural members supporting the machine are away from these critical speeds.

Initial Foundation Layout. The preliminary foundation layout considerations, including basemat size, column and wall sizes, and tabletop geometries, can be found in Chapter 3. For elevated space-frame pedestal foundations, TG manufacturers provide the initial sizes of tabletop members and columns, to accommodate machine installation and pipe routing. Any changes in size required by the design engineer should be discussed with the responsible mechanical engineer and the machine manufacturer.

7.2.2 Determine Dynamic Design Loads and Load Combinations

Chapter 4 provides a list and descriptions of design loads and load combinations for the static analysis of TG foundations. This section focuses on specific loads and load combinations required for dynamic serviceability analysis only. Design loads used for dynamic serviceability analysis are called rotor dynamic unbalance loads. Before a TG system starts its actual operation, TG rotors are balanced to low levels of residual unbalance. The balance quality grade (G) for gas and steam turbines is G2.5, with an $e\omega$ product of 2.5 mm/s (0.1 in./s) per ISO 1940-1. The unbalance loads at each bearing are then determined using $F = me\omega^2$ (see Chapter 4, Section 4.3.2). The forces imposed on the foundation by initial rotor unbalance are not significant. However, the unbalance loads may become appreciable resulting from long-term degradation of the rotor or transient events such as thermal shocks. In some cases, a TG system may operate for an extended period of time with a higher rotor unbalance until corrections can be made. Many TG manufacturers consider a balance quality grade one level higher (G6.3) with an $e\omega$ product of 6.3 mm/s (0.25 in./s) when determining the unbalance loads for the dynamic analysis. The foundation design engineer should check with the TG manufacturer to understand the rotor dynamic unbalance loads provided for the dynamic serviceability analysis and the corresponding foundation vibration acceptance criteria.

The following are the basic loads to be applied at the shaft bearing locations of a TG system for dynamic serviceability analysis:

- Turbine unbalance loads; and
- Generator unbalance loads.

If the TG manufacturer provides only magnitudes of unbalance forces, the basic load combinations should include in-phase and out-of-phase conditions, or a square root of the sum of the squares (SRSS) combination of the foundation responses to the unbalance loads from each rotor/shaft portion acting individually. If the TG manufacturer provides unbalance forces with specified phase angles at each bearing location, the foundation design engineer should use this information in developing the forcing functions. Note that a more conservative method is to sum the foundation responses from the contribution of each unbalance force; phase differences then need not be considered. Load factors used in the load combinations for dynamic serviceability analysis should be 1.0, unless the TG manufacturer has special requirements.

7.2.3 Calculate Dynamic Soil-Pile Impedance

Dynamic soil or pile impedances are frequency-dependent. Detailed information on determining the soil-pile impedance values can be found in Chapter 5 of this book and in ACI 351.3R.

The supporting media profile underneath the TG foundation may deviate from those at boring locations because soil properties are subject to inherent variations. Therefore, it is difficult to accurately predict the actual dynamic soil-pile impedance for each TG foundation. In order to account for this uncertainty, two methods are recommended below:

1. U.S. practice (Arya et al. 1979) recommends using three sets of soil conditions: best estimate, lower bound, and upper bound. The best estimate is the soil or pile spring element properties (stiffness and damping) that result from the average soil properties. The upper bound and lower bound can be taken as +/− 30% to 50% of the best estimate (See Eq. 5-39. Also see ASCE 7, Section 12.13) obtained from direct field and laboratory tests and from correlated methods, respectively. The range used between the best estimate properties and the upper and lower bound values depends on the methods used to calculate the dynamic soil properties.
2. Instead of varying the spring element properties, an alternative is to vary the excitation frequency. European practice (BS EN 1991-3: 2006) recommends using a range of +/− 5% of the excitation frequency with a constant excitation force. This frequency range corresponds to a variation of +/− 10% of the overall system stiffness. Some TG manufacturers recommend using a range of +/− 10% of the excitation frequency, which corresponds to a variation of +/− 20% of the overall system stiffness.

These methods mentioned are generic. For a specific project, the actual range of soil or pile properties and excitation frequency should be determined based on the geotechnical consultant's recommendations and the TG manufacturer's requirements.

7.2.4 Create Dynamic FE Models of TG Foundation System

Chapter 6 provides details on FE modeling of the TG foundation for the dynamic analysis (e.g., selection of element type and sizes, modeling of damping, and boundary conditions). This section specifically addresses the dynamic forcing functions used in the dynamic FE models. In order to simulate excitation forces using dynamic forcing functions, the following factors should be included in the dynamic analysis of the TG foundation:

- *Range of Loading Frequencies.* While the dynamic response of the TG foundation at the operating speed is of particular interest, machine speeds in the frequency range of 80% to 120% of the operating speed, which encompass the startup and shutdown conditions, are also important. Thus it is desirable to construct a frequency response curve as shown in Fig. 7-1 for design quantities (e.g., displacements, velocities).
- *Phasing of Unbalance Forces.* A set of design values of the rotor unbalance force at the operating frequency should be obtained from the TG manufacturer at each bearing of the TG machine set, and the unbalance forces should be assumed to act simultaneously at all bearings.

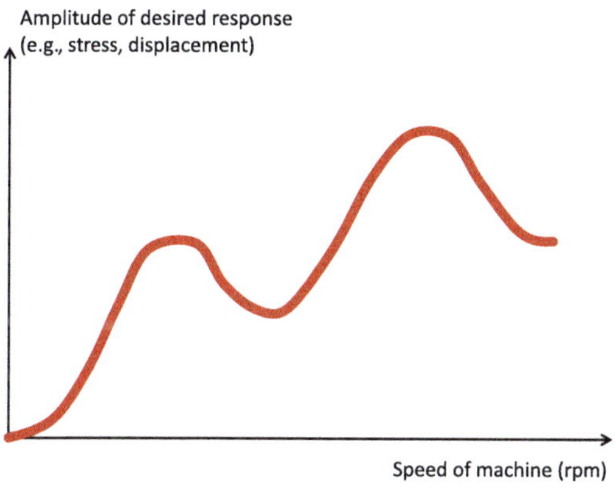

Fig. 7-1. Response spectrum for loads applied to foundation

The unbalance forces should be arranged to yield the most probable maximum response. In general, the arrangement of unbalance forces that causes maximum bearing displacement will not be the same as the arrangement that causes maximum bearing reactions. Also, the assumption that all the unbalance forces are in phase or aligned may not lead to a maximum displacement.

As discussed in Section 7.2.2, if the TG manufacturer only provides magnitudes of unbalance forces, the forces can be defined as vertical and lateral forces and applied at each bearing point separately. As a minimum, the basic load combinations should include in-phase and out-of-phase conditions. If the TG manufacturer provides unbalance forces with specified phase angles at each bearing location, design engineers can create more realistic forcing functions and apply them at each bearing point. In this case, no in-phase and out-of-phase conditions are required.

A typical dynamic forcing function can be written as follows:

$$VH_{i-j}\,(\omega t) = Amp_{i-j} \times sin\,(\omega t + \alpha_{i-j}) \tag{7-1}$$

where

VH = vertical or horizontal dynamic forcing function with a force unit (kN or kips);

i = the ith excitation frequency;

j = the jth TG bearing;

Amp = the maximum unbalance force amplitude with a force unit (kN or kips);

ω = operating annular frequency;

t = passing time;

α = phase angle.

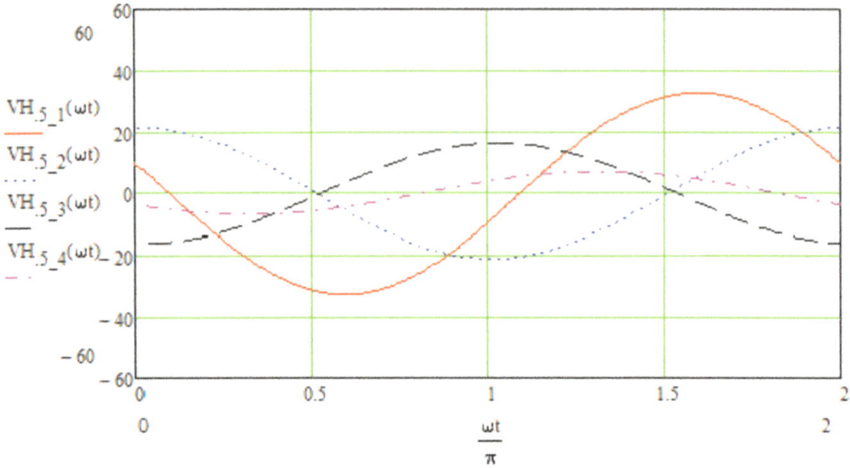

Fig. 7-2. Example of dynamic forcing functions at four bearings

Fig. 7-2 shows an example of dynamic forcing functions at four TG bearings based on excitation frequencies and phase angles specified by a TG manufacturer. As illustrated in the figure, the excitation force at Bearing 4 has the smallest amplitude and a phase angle close to excitation force at Bearing 3. The excitation force at Bearing 2 is almost completely out of phase with excitation force at Bearing 3.

7.2.5 Perform Dynamic Serviceability Analyses

Once the model has been created, a dynamic analysis may be performed to determine the vibration displacements and velocities at the TG bearing locations. These values are necessary to determine whether the foundation will meet the specified dynamic acceptance criteria imposed by the TG manufacturer.

The dynamic analysis of a TG foundation system is performed in two stages:

1. Frequency Analysis, a.k.a. "Modal/Eigenvalue Analysis," to determine the natural frequencies and mode shapes of the TG machine foundation system. Note that a modal or eigenvalue analysis is one common way to perform frequency analysis. It can also be performed using the frequency sweep analysis shown in Fig. 7-4.
2. Forced Vibration Analysis to calculate the TG machine foundation system responses induced by the dynamic forces for the range of frequencies of interest.

In addition to these two previous stages, some TG manufacturers may require a member local frequency check for certain structural members directly supporting the TG machine set. This prevents local foundation resonances with several fundamental TG machine critical speeds, which are calculated by the TG manufacturer using in-house rotor dynamic FE models.

A **frequency analysis** will provide the dynamic characteristics of the TG system, for example, natural frequencies, mode shapes, and modal participating mass ratios or "mass participation factors" (MPF). The computed natural frequencies can be used for the resonance check of the machine foundation system. Most TG manufacturers require that a TG foundation have no fundamental or major frequencies within an exclusion range of +/− 20% of the operating frequency. The highest calculated frequency should be at least 20% higher than the turbine operating frequency.

A TG foundation is a massive and complex structural system that has many mode shapes, global and local. The modal shapes and behaviors can be used to identify potential global and local foundation vibration issues.

The frequencies and mode shapes can be utilized by the modal superposition method as the basis to perform forced vibration analysis. In many cases, six fundamental frequencies are calculated consistent with six degrees of freedom of the overall machine foundation system. Any or all six frequencies may be compared with the excitation frequency when checking for resonance conditions.

ACI 351.3R provides additional information and discussions on frequency analysis.

For a TG foundation, even though frequency analysis provides a wide spectrum of structural natural frequencies, which can be used to ensure that there are no major modes (with large values of MPF) close to the machine operating frequency and/or critical speeds, it is usually not sufficient to conclude that the TG foundation will not have excessive vibrations. This is because some frequencies will exist within a close range of the operating speed, causing dynamic excitations in the foundation. The foundation's response to such excitations will be overlooked unless a detailed forced vibration and/or frequency sweep analysis is performed. Moreover, many structural analysis codes calculate MPF on the assumption that the load is a ground motion. Therefore, such MPF are not necessarily meaningful when the load corresponds to forcing functions applied at a few locations. For additional details refer to Chopra (1995, Sections 12.9, 12.10, and 12.11).

A **forced vibration analysis** (also "harmonic or steady-state analysis") calculates the peak steady-state foundation responses (e.g., displacements, velocities, accelerations, forces) for harmonic dynamic unbalance loads applied at each forcing frequency. This analysis is a frequency domain analysis. The results can be compared with the defined criteria or recommended allowable values for a specific condition. It is recommended to use a steady-state analysis or equivalent periodic time-history analysis for forced vibration analysis based on current industry common practices. If the TG manufacturer provides only magnitudes of unbalance loads, there are two methods to determine the maximum vibration level of TG foundation:

1. One method that can be used to compute the maximum vibration displacements, velocities, and accelerations is to specify a set of load combinations including in-phase and out-of-phase conditions. The number of load combinations should be determined according to the requirements of TG manufacturers.
2. An alternate method is to analyze for the unbalance force at each bearing of the TG set. A conservative, upper limit of the response for each frequency of the unbalance force is the sum of the absolute values from each load acting separately. Practically, the maximum response for each frequency of the unbalance force can be determined as the square root of the sum of the squares of the values for each load acting separately.

Fig. 7-3 shows an example of results from a forced vibration analysis. The results plotted are zero-to-peak vibration displacements at each bearing location for a frequency range of 54 to 69 Hz (from –10% to +15% of the operating frequency of 60 Hz).

A TG manufacturer may provide the machine damped critical speeds or frequencies, and require the TG foundation support system frequencies to lie outside the range of these critical speeds. In such cases, it is necessary to perform a frequency sweep analysis on selected members to determine the member's local resonance frequencies are away from these critical speeds.

A **frequency sweep analysis** is similar to a force vibration analysis in that both use harmonic loads over a frequency range. The difference is that a forced vibration analysis predicts actual steady-state responses of the foundation for each frequency of interest, whereas a frequency sweep analysis involves anticipating and selecting local vibration point(s) and applying a set of unit sinusoidal forces at that point in an anticipated vibration direction. The magnitude of this force is a unit constant force, but the frequency varies. A dynamic amplification factor (DAF) is calculated for each frequency as the response amplitude divided by the static response (i.e., at zero frequency). A plot of DAF versus frequency for selected points on the foundation can be generated. A peak in the plot indicates a possible local resonance in the selected point and the member containing that point in the anticipated vibration direction. The frequencies at which the peaks occur can be compared with the machine critical speeds to ensure that no local member resonances will occur at the machine critical speeds.

Note that finding resonance local frequencies as described previously depends on how small a frequency step is used in the sweep analysis. Local

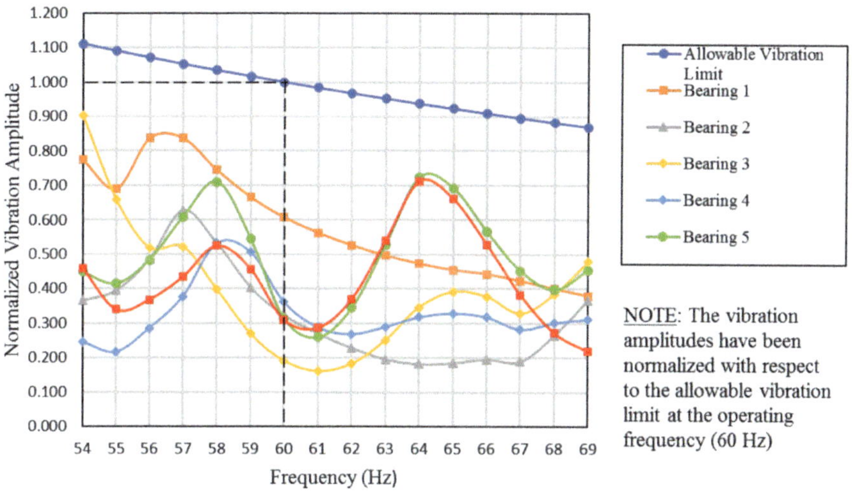

Fig. 7-3. Example of vibration amplitudes from a forced vibration analysis, 54 Hz–69 Hz

peaks could be easily missed if a large frequency interval is used, particularly in lower frequency ranges for structural elements in TG foundations. Alternatively, interpolation in the frequency domain can be used to reconstruct the frequency response function (FRF) of interest. Detailed discussion can be found in Ushijima and Kausel (1981), Lysmer et al. (1999), and Fok et al. (1986).

Fig. 7-4 shows an example of a frequency sweep analysis. In this example, a local peak occurs at 4 Hz, indicating that a local resonance would occur at the selected point (in this case, a point located on a foundation tabletop beam member), if a machine component near that member has a critical frequency of 4 Hz.

Based on results of the dynamic analyses discussed here, the proportions of the foundation (shape, width, length, and depth), foundation embedment depth, and pile arrangement (if applicable) are modified by the design engineer and the modified structure is reanalyzed and reevaluated. This process may take a number of iterations until the project specific vibration acceptance criteria are completely satisfied.

7.3 STATIC SERVICEABILITY ANALYSIS

The preliminary TG foundation layout should also be evaluated with a static serviceability analysis in addition to the dynamic analysis described in Section 7.2.

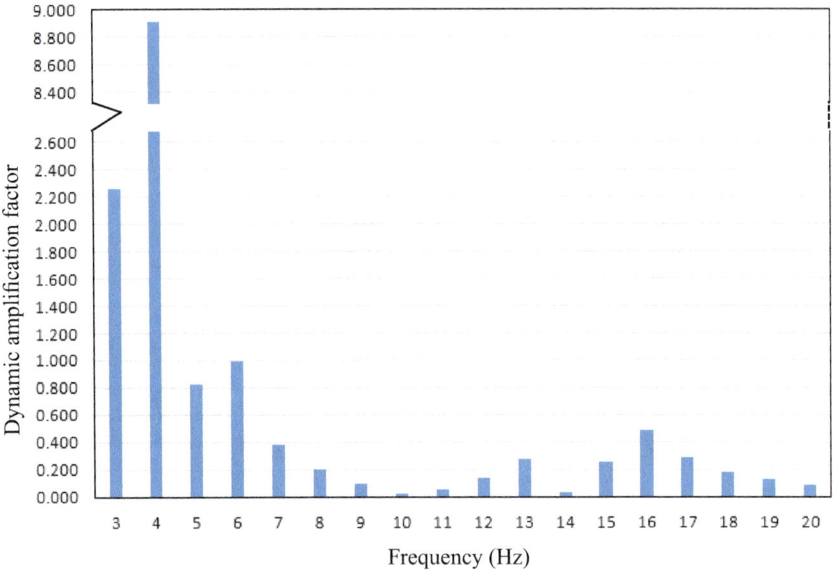

Note: At each frequency, the dynamic amplification factor is calculated as the ratio of the dynamic amplitude due to a unit dynamic excitation, to the static deflection due to a unit static load, both applied at a location of interest of a structural member.

Fig. 7-4. Example plot of a frequency sweep analysis

A typical procedure for the static serviceability analysis is:

1. Collect static design input and acceptance criteria;
2. Determine static design loads and load combinations;
3. Calculate static soil/pile stiffness;
4. Create static FE models of TG foundation system; and
5. Perform static serviceability analysis.

7.3.1 Collect Static Design Input and Acceptance Criteria

Two types of design input should be collected for static serviceability analysis: geotechnical design information and TG design information. See Chapter 3 for a complete list of information required.

The TG manufacturer will have to provide the static serviceability criteria to the TG foundation designer.

7.3.2 Determine Static Design Loads and Load Combinations

The design loads used for static serviceability analysis are static or equivalent static loads. The following list provides typical design loads that should be included in a static serviceability analysis when applicable:

- Normal operating torque load;
- Condenser vacuum load;
- Thermal expansion/contraction load; and
- Piping loading.

Any deflection calculation that intends to safeguard the shaft alignment should consider only those loads that are applied to the structure after initial alignment is complete. The loads included in the deflection calculation should include all post-alignment static normal operating loads. The condenser vacuum load and normal torque load can cause appreciable foundation deflection subsequent to the initial alignment. Transient foundation loads such as turbine generator thermal friction loads, turbine generator emergency loads, and wind and/or seismic loads need not be included in the static serviceability analysis. Rotating, time-dependent, and bearing unbalance loads need not be considered in the analysis, either. Some TG manufacturers may require evaluating the bearing deflections due to emergency loading conditions. In that case, emergency operating loads should be included in the analysis.

In addition to these loads, some turbine manufacturers may require the following load conditions to be included:

- Equivalent static load due to dynamic load;
- Normal axial thrust load;
- Dead load of machine and foundation; and
- Thermal gradient.

It should be noted that dead load is used to calculate long-term foundation settlement and concrete creep. Permanent foundation deflections and settlements that occur prior to the turbine generator initial alignment do not need to be evaluated in the static serviceability analysis, since these deflections and settlements have already been "zeroed out" during the initial TG alignment.

The effects of thermal gradients within the foundation owing to heat transfer from the equipment and the effects of creep may or may not need to be considered, depending on the TG manufacturer's requirements. Thermal gradients on the TG foundation result from machine operating heat transfer or ambient temperature changes. The values of the differential temperature are usually provided by the TG manufacturer or the civil/structural design criteria for the project.

Load factors used in the load combinations for serviceability analysis should be 1.0, unless the TG manufacturer has special requirements.

7.3.3 Calculate Static Soil-Pile Stiffness

The difference between static and dynamic soil-pile supports in the analytical models lies in the spring element properties. Dynamic spring element properties include both stiffness and damping while static spring constant is only the spring stiffness. In addition, the dynamic soil-pile stiffness is frequency-dependent while the static soil-pile stiffness is not. Static soil-pile stiffness is important for sizing the mat, evaluating pile layout, and checking total and differential settlement. It is not necessary to use a range of stiffness values to cover soil uncertainty, as described in Section 7.2.3 for dynamic analysis. Using a lower stiffness value yields a conservative result.

Unlike dynamic soil stiffness, static soil stiffness is usually calculated using a modulus of soil subgrade reaction multiplied by the foundation basemat area. This method generally leads to a conservative static stiffness. A more realistic way to calculate the static soil stiffness is to compute the soil strain level under the maximum static serviceability load combination. The soil shear modulus can be reduced proportionally from the dynamic (low-strain) shear modulus based on that strain level. The static soil stiffness can then be calculated proportionally from the dynamic soil stiffness.

The static pile stiffness of a single pile (usually provided by the geotechnical engineer) is obtained by dividing the pile capacity by the corresponding pile displacement. Pile group stiffness needs to be reduced based on the ratio of pile spacing to pile diameter.

7.3.4 Create Static FE Models of TG Foundation System

The static model of the TG foundation system can be created independently or by modifying the dynamic model. The static model created can be used for both static serviceability analysis and the strength design. The various aspects of finite element modeling are covered in Chapter 6.

The static model for strength design may be an integrated FE model with foundation basemat and pedestals or two separated FE models: one for the superstructure, with columns and walls fixed at the base, and one for the foundation basemat. If two separate models are used, the design loads are applied to the FE model of the superstructure to determine its responses. The computed reactions at the base of the columns and walls are then used as input forces on the FE model of the basemat.

For static serviceability analysis, the tabletop, columns, and basemat should be incorporated into one single model.

For static analysis, it is recommended that the minimum specified design strength of concrete be used, so that calculated deflections are conservative.

7.3.5 Perform Static Serviceability Analysis

After the static serviceability model has been constructed, static serviceability analysis should be performed to calculate the differential displacements of all the bearings, beam/girder rotation, curvature radius, static stiffness/flexibility, maximum crack width, and differential settlement of TG foundation, as required per TG manufacturer's criteria as well as code requirements. The static serviceability analysis is similar to the analysis for the strength design (Chapter 8) with the following differences:

1. Accidental loads or extreme environmental loads are not considered for static serviceability analysis. Emergency operation loads typically are not required, but can be included per the requirements of machine manufacturers. The typical loads used for the static serviceability analysis include normal torque load, piping load, and thermal load. The detailed discussions can be found in Section 7.3.2.
2. The load factors in the load combinations should be 1.0 or less unless the TG manufacturer has special requirements.
3. The static serviceability analysis is a working-stress level analysis.
4. The typical output of static serviceability analysis is displacements. Resultant moment or stress will be used only for crack width calculation, if specifically required by the TG manufacturer.

7.4 SERVICEABILITY ACCEPTANCE CRITERIA

The general purposes of TG foundations are

1. To serve as a stable and secure base for the TG machine set during erection and alignment;
2. To provide a base with adequate rigidity so that the alignment of the TG machine set may be maintained during both normal and emergency operating conditions;
3. To transmit the dead and live loads from the TG machine set to the supporting media (soil or piles) in such a way that no detrimental effects occur in the machinery, the TG foundation, or the supporting media (soil or piles); and
4. To maintain an acceptable level of vibration and minimize the dynamic effects from the machinery that might otherwise cause excessive wear and damage to the TG machinery, damage to the TG foundation, or discomfort to the operating staff.

To meet these objectives, appropriate foundation performance acceptance criteria, for both dynamic and static considerations, should be established for the design of the TG foundation.

7.4.1 Dynamic Acceptance Criteria

Historically, the dynamic coupling between the turbine generator and the foundation has not been a serious consideration for the TG foundation design engineer. The traditionally massive and rigid conventional concrete foundations resulted in limited interaction between the foundation and the power-generating equipment. Consequently, dynamic response criteria were not required, and the turbine design engineer and the TG foundation design engineer could work without regard to the dynamic characteristics of the other's design.

Turbine generators have grown in rating and size. Coupled with a trend toward more flexible foundations, increased attention has been given to the dynamic characteristics of the turbine generator foundation system. The concerns of the turbine manufacturer are related to the reliability of the machine and the impact of the unit on the TG foundation. Small amounts of inherent unbalance in the turbine generator can be amplified through resonance with the foundation to produce excessively large bearing forces. Large relative bearing displacements could also produce large bearing forces and high rotor bending stresses. These adverse conditions could affect the operability of the unit or lead to a forced outage. In addition, excessive TG foundation response could be judged uncomfortable for humans, leading to unsatisfied clients. For more information on human tolerance vibration levels, refer to ACI 351.3R.

In order to ensure satisfactory and reliable operation of the turbine generator and to provide a working environment around the unit, TG manufacturers specify dynamic response characteristics. These characteristics have taken the form of a maximum TG foundation response under normal conditions or maximum levels of other vibratory characteristics.

One vibratory characteristic that has been suggested by ISO 20816-2 for describing the vibratory state of a machine is known as the "vibration severity." The vibration severity is defined as the root-mean-square value of the vibration velocity.

The following are generic dynamic acceptance criteria, which provide readers a general vision of what to look for in the dynamic analysis for a TG foundation:

1. Frequency Resonance Exclusion Criteria—The natural frequencies of the foundation system should be situated such as to avoid resonance with the machine operating speed. Per ACI 351.3R, a typical requirement is that the natural frequency be 20% to 33% away from the operating speed. However, the frequency criteria for a TG foundation should be based on the project specifications and the turbine manufacturer's requirements.
2. Forced Vibrating Response Criteria—The amplitudes of the foundation response at the bearing housing or pedestal locations must satisfy the requirement of ISO 20816-2, Table A.1, Zone Boundary A/B for new units operating under normal operation. For evaluation of TG foundation vibration response on existing units or for

emergency conditions of new units, less stringent criteria may be used.

3. Other Dynamic Criteria—Some TG manufacturers may also have foundation dynamic stiffness/flexibility criteria. Different from a traditional structural static stiffness, a structural dynamic stiffness is not only a function of the location and direction where the stiffness is measured, but also a function of the dynamic forcing frequency.

If required, a foundation dynamic stiffness should be calculated, at each location (bearings or at machine foundation interfaces) and in each direction stipulated by the machine manufacturer, one at a time, as follows:

1. Apply a unit concentrated load ("F") at the location in the direction of interest. The load should be a dynamic excitation at the frequency of interest.
2. Obtain the peak vibration amplitude ("d") at the loading point in the same direction where the unit load is applied.
3. Calculate the dynamic stiffness as "F/d" for the excitation frequency.

7.4.2 Static Acceptance Criteria

Turbine generator foundation design engineers use foundation design criteria based on static deflections of the foundation under a specifically identified set of loading conditions. In the design of TG foundations, deflections are first and foremost a functional requirement. TG manufacturers have many reasons to limit the deflection of the TG foundation to stipulated limits.

A typical shaft is supported by a number of bearings. Excessive differential deflections between adjacent bearings can induce large stresses in the shaft. In addition to large shaft stresses, differential deflection between adjacent bearings may result in overloading some of the bearings while other bearings are underloaded. Since each bearing is specifically designed for a range of minimum and maximum loads to optimize the bearing performance, TG manufacturers recommend that the integrity of the bearings not be jeopardized by over- or underloading any bearing beyond an acceptable margin.

For all normal operating conditions, the minimum running clearances between the stationary and rotating elements of the machine must be maintained in the turbine interstage and shaft gland steam seals and in the generator hydrogen seals. One way to achieve smooth operation of a machine is to limit the total deflection, and in turn the differential deflection, of various structural components of the TG foundation.

In order to achieve these desired objectives, the TG manufacturers may use a variety of methods to specify the deflection limits of TG foundations.

TG manufacturers can produce TG machine sets that can have output of more than 1,000 MW. The foundation that supports large TG machinery becomes increasingly complex, not only in geometries, but also in more stringent design requirements that do not exist for smaller TG machine sets (<50 MW). To properly control the deflections of foundations supporting large modern TG sets (>50 MW), TG manufacturers impose foundation deflection limits.

Misalignment Tolerance Matrix (MTM) Criteria. One such criterion is the misalignment tolerance matrix (MTM) method. The MTM defines the percentage of available misalignment that is consumed at a given bearing due to displacement at another bearing. It can be described as an influence coefficient approach.

This criterion involves defining the amount of misalignment that a turbine generator rotor system can withstand without imposing unacceptable loads on the rotor bearings or unacceptable bending stresses in the rotor system.

The TG foundation design engineer calculates the deflections that may result from various loads that are applied to the foundation after the initial turbine alignment. The design engineer then uses the influence coefficients provided by the TG manufacturer to determine whether the TG foundation's response is likely to adversely affect the turbine generator operation or reliability.

The influence coefficient is the factor by which a deflection is multiplied in order to obtain the response at a specific location. The response may be the load at a bearing support point, the bending stress in the shaft, or a normalized general force. These responses are compared against a given set of allowables (minimum and/or maximum bearing load, maximum stress, etc.) to ensure that the deflection of the TG foundation does not adversely affect the operation and reliability of the TG machine.

As an example, consider the single-span, propped cantilever beam shown in Fig. 7-5. If the beam is straight and has no initial support deflection, the response at each of the locations shown in Fig. 7-5 is R'_1, R'_2, and R'_3. These responses could be beam member forces, support reactions, beam deflections, or stresses. Due to support settlement or an initial temperature deflection, the three points will displace vertically upward ($+y$) or downward ($-y$). These deflections are labeled y_1, y_2, and y_3. These responses of the beam system are a function of the beam geometry, support stiffness, and distances between points of interest. The particular response due to a displacement can be expressed in terms of a series of influence coefficients (C), which are unique for a specific beam and a particular response. For example, for a given beam, there is a unique set of three coefficients to describe the bending stress at Location 2 due to a deflection at each of the three locations. This can be expressed for Location 2 as

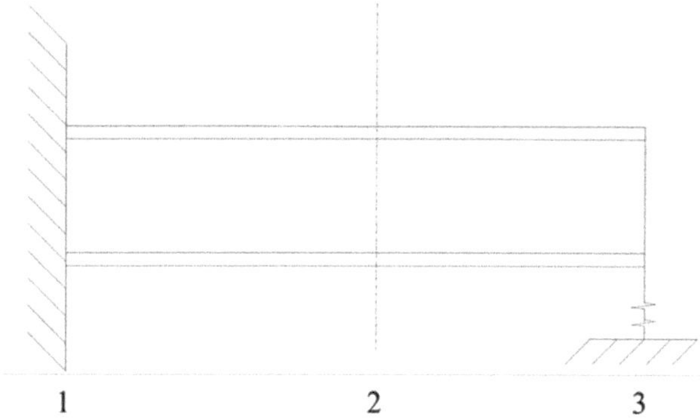

Fig. 7-5. Propped cantilever beam

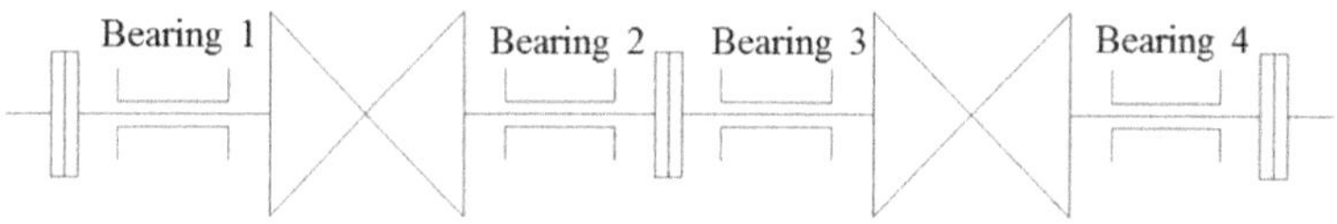

Fig. 7-6. Four-bearing turbine shaft system

$$R_2 = R_2' + C_{21}y_1 + C_{22}\,y_2 + C_{23}\,y_3 \tag{7-2}$$

where

R_2' = the initial response at Location 2,

C_{2j} = the change in response at Location 2 per unit of vertical deflection of Location j, and

y_i = the vertical deflection of Location i from the initial position.

This response may be a deflection, beam bending or shearing stress, support reactions, or beam member forces (moment, shear), depending on the response of interest. The position of the point of interest as well as the required response will determine the sign of the quantity, C_{21}, C_{22}, or C_{23}.

A specific example of the influence coefficient approach for turbine generator foundation design follows. Fig. 7-6 shows a simplified line drawing of the first two casings of a turbine generator with a shaft system supported by four bearings. The rotors are precisely aligned with one another in the cold condition, and the couplings are bolted together. The loadings on the four bearings in Fig. 7-6 would be P_1', P_2', P_3', and P_4', provided the machine could run with undisturbed shaft alignment. During normal operation, however, all the bearings are displaced vertically upward ($+y$) or downward ($-y$) due to TG foundation deflections. The various amounts by which the bearings are loaded and unloaded as a result of such shaft misalignment depend upon the design of the bearings, the configuration of the shaft system, and the distances between the bearings. The influence coefficients indicate the amount by which the loading on a bearing changes per inch (mm) of bearing support displacement. The load on Bearing 2 as a result of shaft misalignment is

$$P_2 = P_2' + (-C_{21})y_1 + (+C_{22})y_2 + (-C_{23})y_3 + (+C_{24})y_4 \quad (7\text{-}3)$$

where

P_2' = calculated loading of Bearing 2 with correctly aligned shaft system in psi (kN/m^2),

C_{2j} = change in loading on Bearing 2 per inch of vertical deflection of Bearing j in psi/in. (kN/m^2/mm), and

y_i = vertical deflection of Bearing i from the position of initial correct shaft alignment in in. (mm).

The influence factor C_{22} will be positive and have the largest value, reflecting the fact that an upward displacement of Bearing 2 strongly increases the load on the same bearing, whereas a downward displacement results in a correspondingly large decrease in load. The negative factors C_{21} and C_{23} denote that upward displacement of Bearings 1 and 3 tends to unload Bearing 2. The value of C_{23} will, of course, be considerably larger than C_{21} because of its proximity to Bearing 2. The effect of other bearing displacements on the loading of Bearing 2 becomes smaller with increasing distance from Bearing 2, so that the factor C_{24} will have the smallest value.

The vertical and horizontal coefficients at each of the bearings are required from the TG manufacturer. After the results of MTM analysis are obtained, the responses are compared to the allowable values given by the TG manufacturer. The units as well as the locations of the required displacements are specified by the TG manufacturer.

Some forms of shaft misalignment can have an adverse effect on the bearing loadings but not increase the bending stresses in the shaft system. It is equally possible for some shaft systems to be vulnerable at particular locations to high bending stresses that do not result in any substantial alteration of the loads on individual shaft bearings. When the combined effect of the foundation deflections is found to exceed the allowable response at a particular point, it should be recognized that a design change at another location can be a feasible solution.

Some TG manufacturers provide one set of influence coefficients to determine the effects of foundation displacement on the bearing loads and a different set of coefficients to investigate the effects on the shaft bending stresses. Other manufacturers combine the effects of bearing loads and shaft stresses into a single set of coefficients. In either case, the static serviceability analysis by the foundation design engineer is the same.

Other Static Serviceability Deflection Criteria. TG manufacturers may also impose other deflection criteria as follows:

1. Differential radial displacement between adjacent bearings;
2. Foundation basemat differential settlements;
3. Relative top surface rotation of specific piers;
4. Relative displacements of any three adjacent foundation supports; and
5. Concrete crack width of top deck girders.

Note that the design of TG foundations is typically governed by stiffness criteria; as such, stress levels in concrete sections and reinforcement are relatively low compared to those in building and other types of structures. Therefore, fatigue from cyclic loading in TG foundations is usually not a concern.

It is essential that the TG foundation design engineer have a correct understanding of all the serviceability criteria, both static and dynamic, imposed by the TG manufacturers. Coordination with the TG manufacturers is the key to a successful design if the foundation design engineer has difficulties meeting certain criteria.

REFERENCES

ACI (American Concrete Institute). (2004. Reapproved 2011). "Foundations for dynamic equipment." *ACI 351.3R-04*, Detroit.

ACI. (1976). "Testing hardened concrete: non-destructive methods." *ACI Monograph No. 9*, Detroit.

ACI. (2014). "Building code requirements for reinforced concrete." *ACI 318-14*, Detroit.

Arya, S., O'Neill, M., and Pincus, G. (1979). *Design of structures and foundations for vibrating machines*. Gulf Publishing, Houston.

ASCE. (2010). "Minimum design loads for buildings and other structures." *ASCE/SEI 7-10*, Reston, VA.

BS EN. (2006). "Eurocode 1: actions on structures, part 3: actions induced by cranes and machinery." *BS EN 1991-3*, London.

Chopra, A. K. (1995). *Dynamics of structures: theory and applications to earthquake engineering*. Prentice Hall, Englewood Cliffs, NJ.

Fok, K., Hatt, J. F., Chopra, A. K., and University of California, Berkeley. (1986). *EACD-3D: a computer program for three-dimensional earthquake analysis of concrete dams*. Earthquake Engineering Research Center, University of California, Berkeley.

ISO (Internation Organization for Standardization). (2003). "Mechanical vibration—Balance quality requirements for rotors in a constant (rigid) state—Part 1: Specification and verification of balance tolerances." 2nd Ed., *ISO 1940-1*, Geneva, Switzerland.

ISO. (2017). "Mechanical vibration—Measurement and evaluation of machine vibration—Part 2: Land-based gas turbines, steam turbines and generators in excess of 40 MW, with fluid-film bearings and rated speeds of 1,500 r/min, 1,800 r/min, 3,000 r/min and 3,600 r/min." *ISO 20816-2*, Geneva, Switzerland.

Lysmer, J., et al. (1999). *SASSI 2000 theory manual: A system for analysis of soil-structure interaction.*

Ushijima, R., and Kausel, E. (1981). "Two efficient schemes for dynamic analyses with finite elements." *Nuclear Engineering and Design* 66(1), 141–146, ISSN0029-5493.

CHAPTER 8
STRENGTH AND STABILITY DESIGN

8.1 INTRODUCTION

The strength design of TG foundations should follow requirements of the building codes specified in the project design criteria, as well as applicable ACI requirements. However, TG foundations have some unique structural characteristics that require special considerations and details in strength design.

This chapter presents the strength design criteria and procedures for the design of concrete TG foundations. The stability design considerations are also discussed.

8.2 LOAD COMBINATIONS FOR STRENGTH DESIGN

The strength design method should be used to design all structural components of reinforced concrete TG foundations.

Factored load combinations for strength design should follow the applicable building codes. In addition to the general load combinations, special load combinations required or recommended by the TG machine manufacturer should be considered. Loads to be used in load combinations are described in Chapter 4. Section 4.9 discusses load combination considerations for TG foundation designs.

8.3 SEISMIC LOAD AND DUCTILE DESIGN CONSIDERATIONS

Seismic loads are defined by the building code specified in the project civil/structural design criteria. Key seismic design parameters, such as S_{DS}, S_{D1}, soil type, and seismic design category (SDC), are usually provided in this document.

According to Section 11.7 of ASCE 7, TG foundations assigned to SDC A need only comply with the requirements of ASCE 7 Section 1.4. TG equipment systems in SDC A are exempt from seismic design requirements. For TG foundations in SDC B to F, the equivalent lateral force analysis procedure is permitted. The seismic response coefficient C_s and the response modification factor R can be determined based on the following design considerations.

For a rigid TG foundation system with a fundamental period less than 0.06 seconds, it is acceptable to use $C_s = 0.3\ S_{DS}\ I_e$ per Section 15.4.2 of ASCE 7. Note that S_{DS} is the site design response acceleration as determined from Section 11.4.4 of ASCE 7, while I_e is 1.25 for Risk Category III and 1.5 for Risk Category IV, respectively. Some low-profile CTG, axial exhaust, or side exhaust STG with higher soil or pile stiffness may be qualified as rigid foundations for the fundamental horizontal modes.

For TG foundations with a fundamental period not less than 0.06 seconds, the seismic response coefficient C_s can be determined per Sections 12.8.1 and 15.4.1 of ASCE 7. In these procedures, the design engineer needs to select the response modification factor R for the TG foundation, which reflects the capabilities of the structure to absorb and dissipate earthquake energy during such an event.

However, the current practice in selecting an appropriate R-value for seismic design of TG foundations varies, depending on how the structure type classification per ASCE 7 is interpreted, seismic severity, and past experiences of the TG foundation design engineer.

Some practitioners classify the elevated space-frame TG foundations as "nonbuilding structures similar to buildings." The corresponding seismic coefficients, with their usage applicability and limitations, are listed in Table 15.4-1 of ASCE 7. Accordingly, the design engineer can select an R-value of 0.8 from Table 15.4-1, classifying the TG pedestal as an ordinary reinforced concrete moment frame (OMF), to avoid ductile detailing requirements stipulated in ACI 318. However, using an R-value of 0.8 will inevitably result in higher seismic demands on the foundation, which may not necessarily cause reinforcement issues in the superstructure, but will very likely cause lateral pile capacity issues for pile-supported foundations, or soil-bearing capacity issues for soil-supported foundations, or even excessive sliding under high seismic loading. For pile-supported foundations, increasing the number of piles alone may not increase the total lateral capacity sufficiently, because individual pile lateral capacity becomes lower as piles are spaced more closely together. As a result, additional piles with a larger basemat footprint may be required, possibly leading to new interface issues with nearby structures.

To avoid issues described previously, alternatively, the design engineer more intends to select an *R*-value of 3.0, and follow ACI 318 to meet the seismic detailing requirements for intermediate moment frames (IMF). Note that per ASCE 7, for the OMF and IMF, the structural height limit is NL (no limit) for SDC B and C, and 50 ft for SDC D to F, respectively. However, it is the opinion of this task committee that the 50-ft height limitation is too restrictive for a TG pedestal foundation, whose design is more governed by stiffness and vibration criteria than by seismic strength requirements. If a foundation is proportioned per the rules of thumb on weight ratios, column axial compressive stress range, and h/r ratio, as recommended in Chapter 3, the 50-ft height limitation, can be 60 ft for SDC D to F.

Contrary to the previous practice, other practitioners treat TG foundations as "nonbuilding structures not similar to buildings." Then, per Table 15.4-2 of ASCE 7, an *R*-value of 1.25 can be selected for "all other self-supporting structures, tanks, or vessels not covered above or by reference standards that are similar to buildings." Or, an *R*-value of 2.0 can be selected if elevated space-frame TG foundations are treated as "inverted pendulum type structures." A survey of major engineering design firms shows that, in various practices, *R*-values of 1.0, 1.25, and 2.0 have been used without providing seismic detailing required by ACI 318. Among considerations, one major defense argument for doing so has been that TG foundation designs are governed by stiffness and vibration limits, not by seismic strength requirements. As a result, a TG pedestal foundation is usually oversized, such that its structural responses under seismic loads are more or less elastic than plastic. Therefore the detailing requirements may not be necessary for these oversized structural components.

While acknowledging the fact that different *R*-values have been used in current practice, this task committee recommends using either $R = 0.8$ or $R = 3.0$, and following the appropriate detailing requirements of ACI 318, since there has not been enough research and evidence to substantiate other *R*-values. A decision to use other values should be based on careful considerations of specific project engineering requirements, as well as close coordination with local building officials, as required. In this regard, this Committee expects that future ASCE 7 updates will provide clearer provision(s) on nonbuilding foundation structures for large vibrating equipment.

Besides the superstructure, the seismic inertia effect of the TG basemat supporting the superstructure should also be considered. Section 12.2.3.2 of ASCE 7 provides the details for two-stage analysis procedure. A typical TG basemat may be considered as a rigid structure for this purpose, regardless of whether it is above or below grade. ASCE 7 recommends $C_s = 0.3\ S_{DS}\ I_e$

as the lateral seismic coefficient for rigid structures. For the purpose of strength design, both soil-supported and pile-supported foundation mats are treated as rigid structures under the effect of horizontal seismic excitation. In the latter case, the interaction effects between the pile and surrounding soil medium are ignored.

The seismic loads of the TG machines should be determined treating the machines as an integral part of the foundation. TG manufacturers may provide machine seismic loads, which design engineers should verify.

For machine anchorage designs, the design engineer should follow ASCE 7, Chapter 13.

8.4 REDUNDANCY FACTOR (ρ) AND OVERSTRENGTH FACTOR (Ω_0)

In accordance with ASCE 7, Section 12.4, redundancy and overstrength factors must be used to create a new set of design load combinations for all structural members. These provisions are applicable for buildings and nonbuilding structures per ASCE 7, Sections 15.4.1, 15.5.1, and 15.6.

Per ASCE 7, Section 12.3.4, the value of the redundancy factor (ρ) for TG foundations can be taken as 1.0. TG foundations are "nonbuilding structures that are not similar to buildings" for block type foundations, or meet the requirements for moment frames described in ASCE 7, Table 12.3-3, for elevated space frame type foundations.

ASCE 7 indicates that where specifically required, seismic load effects shall be modified to account for system overstrength, as set forth in Section 12.4.3. Per Section 18.3.3 of ACI 318, the overstrength factor Ω_0 should be used in the shear reinforcement design of OMF columns. Per Section 18.4 of ACI 318, the Ω_0 factor should be used for the shear reinforcement design of IMF columns while a constant factor of 2 should be used for the shear reinforcement design of IMF beams. The Ω_0 factor is not required for the axial, flexural, and torsional design. As indicated in Section 18.4 of ACI 318, the objective of the requirements in 18.4.2.3 and 18.4.3.1 is to reduce the risk of failure in shear in beams and columns during an earthquake. For anchorage design, ASCE 7 provides a revised version that gives Ω_0 factors for all components in Chapter 13.

8.5 ACCIDENTAL TORSION

ASCE 7, Section 12.8.4.2, requires an accidental torsion to be considered for diaphragms that are not flexible. This requirement accounts for the possible difference between the actual mass locations and those considered in the design.

While the tabletop of an elevated space-frame pedestal foundation may be considered a diaphragm that is not flexible, for a TG foundation all the major masses (e.g., machine weights) are already precisely defined and can be accounted for in a finite element model used for the strength design. Therefore, the accidental torsion defined in ASCE 7 need not be included in the design of TG foundations.

8.6 FE RESULTS FOR STRENGTH DESIGN

TG structural members are sized to satisfy dynamic vibration and static deflection criteria. Therefore, a minimal amount of reinforcement often satisfies the strength requirements. Demands because of load combinations involving seismic loads and/or catastrophic equipment loads govern the reinforcement design of a TG foundation.

Using the static FE model, as described in Chapters 6 and 7, with all applicable design loads and load combinations per building codes as well as the TG machine manufacturer's requirements, the design engineer can perform a static analysis to obtain the design forces for all structural components in a TG foundation.

If beam elements are used to model tabletop members and columns, FE member force outputs (axial force, shear force, moment, and torsion) can be directly used for member designs per ACI 318.

If shell elements are used to model walls and basemat, outputs are usually given in a "per unit length" format (e.g., 300 kip-ft/ft for bending moment in a wall). The design engineer will need to post-process the outputs to obtain the resultant forces along a selected length for reinforcement design. Some computer programs may have tools that help calculate resultant forces at user-defined section cut locations. Also see Section 6.12 for guidelines on averaging FE results.

If solid elements are used to model the whole foundation structure, the outputs are usually stresses, not forces and moments that can be readily used for design. In such cases, the design engineer may have to post-process the stress outputs by doing integrations across the sections of interest to obtain the resultant forces and moments at those sections.

To simplify the design process, enveloping resultant forces and moments can be used for reinforcement design, so long as such an approach does not result in too congested reinforcement arrangements.

8.7 BASEMAT REINFORCEMENT DESIGN

TG foundations have thicker basemats compared to other equipment foundations. The reinforcement design of basemats should follow the applicable ACI requirements. If finite element analysis results are available, the design engineer can obtain the maximum bending moments in both directions, and add the maximum absolute value of the torsional moments to them to determine the required flexural capacity in each direction. The axial force, either tension or compression, in the basemat is typically very small and can be ignored. Some design engineers may add additional reinforcements to resist axial tension.

A simplified approach to design of the basemat is to treat it as a slab supported by the columns and walls, then to use the maximum soil-bearing pressure or maximum pile reactions as input loads to perform hand calculations and determine the maximum moments in the basemat. The flexural reinforcement is provided accordingly.

Typically, no shear reinforcement is necessary in the basemat. Because of its thickness, the basemat concrete provides sufficient shear capacity, larger than the maximum shear demands on the basemat.

A punching shear check should be performed for areas surrounding the columns and piles per ACI 318 provisions.

In foundations thicker than 4 ft (1.2 m), the design engineer may use the minimum reinforcement suggested in ACI 207.2R for mass concrete. Also, Section 7.12.2 of ACI 350 provides an alternative criterion, which states that concrete sections that are at least 24 in. thick may have the minimum shrinkage and temperature reinforcement based on a 12-in. concrete layer at each face. Using this alternative, a typical basemat top (or bottom) reinforcement is 0.5% of the assumed cross section depth (24 in.) in each direction. To minimize concrete cracking in mass concrete, additional minimum spatial steel reinforcement per unit volume of concrete may be required by some turbine manufacturers. More discussion is given in Section 8.11.

8.8 COLUMN, BLOCK, PIER, AND WALL REINFORCEMENT DESIGN

Columns, pedestals, and walls in space-frame pedestal foundations should be designed per the applicable ACI requirements.

The minimum column longitudinal reinforcement should be 0.5% of gross concrete area. Though Section 10.6.1 of ACI 318 requires the minimum longitudinal column reinforcement to be 1% of the gross concrete area, Section 10.3.1 allows the use of a reduction in effective area of up to 50% for the calculation of minimum reinforcement. Therefore, the minimum reinforcement may be as low as 0.5% of the gross area. If the reduced effective area is used for strength calculations and minimum reinforcement, the full gross area should still be used for column stiffness calculations.

The column tie sets should be minimum No. 5 bars with a maximum spacing of 18 in. (450 mm). Tie sets need not go into the tabletop joint for low to moderate seismic zones. For higher seismic zones, ACI 318 seismic detailing requirements should be followed. Note that OMFs do not have special seismic detailing requirements if an *R*-value of 0.8, 1.25, or 2.0 is used.

All longitudinal bars should be laterally supported by ties unless they are 12 in. (300 mm) or closer along the tie to the next laterally supported bar. Lateral support is provided by the corner of a tie with an included angle of not more than 135 degrees.

As an additional check for columns, it is a good practice to ensure that the amount of reinforcement provides a section moment capacity greater than 1.2 M_{cr}, where M_{cr} is the cracking moment capacity based on the concrete modulus of rupture (f_r). This ensures some levels of ductility within the structure.

8.9 TABLETOP REINFORCEMENT DESIGN

Tabletop members should be designed as "beams" per the applicable ACI requirements.

8.9.1 Shear and Torsion

Flexural members of a TG foundation may be subjected to substantial shear and torsion forces. The longitudinal beams supporting the generator stator and the low-pressure turbine exhaust hoods are usually short and heavily loaded, principally along one edge, resulting in relatively high torsion forces. Combined shear and torsion strength of these members should be evaluated by the procedures of ACI 318, taking into account the effects on strength of simultaneous bending moments and axial forces at the section under investigation.

For beams loaded along their top surface, design sections for shear and torsion are taken at a distance "d" from the face of the support unless the shear at the face of the support is substantially different, as in the case where a heavy load or beam reaction is applied within this distance. In this case, the design section should be taken at the face of the support. The distance "d" is defined as the dimension from the compression face of the concrete to the centroid of the tension reinforcement. Shear reinforcement and torsional reinforcement must meet the requirements of ACI 318. The torsional reinforcement should be in the form of closed stirrups.

8.9.2 Deep Beam and Strut-and-Tie Model Method

The beams and girders of a TG foundation tabletop are typically designed as deep beams. Per ACI 318, strain in reinforcement and concrete shall be assumed directly proportional to the distance from the neutral axis, except that, for deep beams, an analysis that considers a nonlinear distribution of strain shall be used. Alternatively, use of the strut-and-tie method (DIN4024-1 1988; Reineck and Novak 2010) is permitted.

If solid elements are used in a static analysis for strength design, the nonlinear distributions of strain in the deep beams will be appropriately accounted for. Therefore, per Section 9.9.1.2 of ACI 318, it is not necessary to use the strut-and-tie method to design such deep members. If beam elements are utilized to obtain the forces and moments for tabletop deep members, the strut-and-tie method should be used. Design engineers can use the traditional beam design method for the initial design of deep beams, and then use the strut-and-tie method to check the reinforcement configurations.

8.9.3 Additional Considerations

The typical minimum reinforcement at beam top and bottom surface should be 0.25% (each) of gross sectional area, and 0.1% (each) at side faces. The typical stirrup configuration is No. 5 closed stirrups minimum, with maximum spacing of 1/5 beam depth but not to exceed 18 in. (450 mm). The minimum tension reinforcement on each face for flexural members is typically controlled by providing at least 1/3 greater than that required by analysis for flexure.

Similar to the column design mentioned in Section 8.8, it is good practice to provide sufficient flexural reinforcement to provide a flexural capacity larger than 1.2 M_{cr}.

Some tabletop layouts may have trenches and block-outs, which make main reinforcement arrangements difficult without bends and cuts. For such layouts, it is recommended to place all main reinforcements beneath the deepest major trench so that they can have a more uniform and regular mesh, making construction much easier. Doing so will reduce the effective moment arm for tabletop member design. However, the flexural strength reductions do not necessarily result in additional reinforcement requirements, since the tabletop member sizes are controlled by stiffness requirements, not by strength requirements. Even if the tabletop layout is uniform (i.e., no large trenches or block-outs), it may still be a good practice to place main reinforcement at least 6 in. below top of concrete, in order to avoid rebar cutting during installation of expansion or epoxy anchors. Note that in this case, secondary reinforcement should be used to minimize the concrete cracking near the surface.

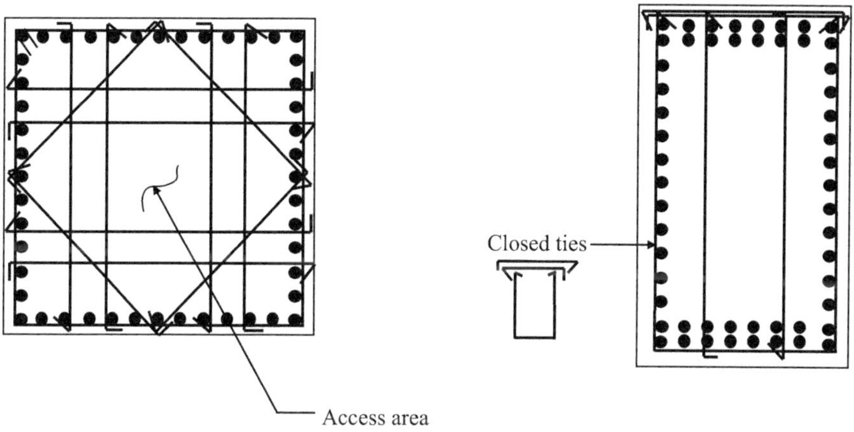

Fig. 8-1. Example of a column tie reinforcing arrangement to allow access inside the rebar cage (left); example of a tabletop beam torsion tie reinforcing (right)

8.10 ADDITIONAL REINFORCEMENT REQUIREMENTS

Following are common practices with respect to reinforcement in TG foundations. Fig. 8-1 shows a typical reinforcement example for a column and a tabletop beam.

- All faces of concrete members should be reinforced with bars spaced not more than 12 in. (300 mm) on center.
- A maximum bar size of No. 11 is preferable. In some cases, No. 14 or No. 18 bars in conjunction with mechanical splicing may be used.
- The clear distance between parallel bars in a layer should be 6 in. (150 mm) or more, unless lateral reinforcement sufficient to prevent bond splitting is provided.
- Splices should be located in compression zones, where possible.
- Careful attention must be given to anchorage and development details because of large reinforcement bar diameters.

8.11 VOLUMETRIC REINFORCEMENTS

While not required by most U.S. codes, some turbine manufacturers or power plant owners require volumetric reinforcement to be provided in TG pedestal foundations per old European practice such as DIN 4024 and CP 2012-1, although these two standards are inactive now.

DIN 4024-1 (1988) is an old German standard for machine foundations. It requires that the mass of reinforcement by base volume should be at least 30 kg/m^3 (or 1.87 lb/ft^3), and part of the reinforcement should be arranged in a spatial configuration. For TG foundations, some design engineers use a minimum reinforcing of 3.1 lb/ft^3 (50 kg/m^3 or 0.64% by volume) for pedestals (no columns) and 1.9 lb/ft^3 (30 kg/m^3 or 0.38% by volume) for foundation basemats. CP 2012-1 is an old British standard for machine foundations. Some machine manufacturers require the use of multiple layers of reinforcing steel spaced in all three directions per CP 2012-1.

Even though the requirements on volumetric reinforcement vary depending on the building code to be followed and/or machine manufacturer requirements, it is generally considered not necessary for TG foundations in the United States, since the foundation stresses are usually very low during normal operations. However, mass pours should be considered with the possible addition of reinforcement as a solution for the thermal issue. Section 6.4 of ACI 207.2R provides the requirements on the minimum reinforcement for mass concrete. The design engineer should collaborate with the machine manufacturer and/or local building officials to make appropriate decisions on specifying volumetric reinforcement.

8.12 PILE CAPACITY CHECK

If a TG foundation sits on piles, the pile group should be laid out and checked by the design engineer to ensure that the maximum demands on the piles are within the pile lateral and axial (tension and compression) capacities provided by the geotechnical engineer.

When checking the pile lateral capacity, the maximum demands come from either seismic lateral loads or catastrophic equipment loads. When determining the seismic lateral demands on the piles, the seismic inertia force of the basemat should be considered, and combined with the superstructure seismic base shear appropriately. The 2011 ASCE Report, "Guidelines for Seismic Evaluation and Design of Petrochemical Facilities," Section 5.3.5.5, provides practical guidelines on how to combine the two seismic shear demands for pile design purposes.

Lateral pile capacities are based on an assumed head deflection that should be checked against project design criteria. Lateral pile capacities provided by the geotechnical engineer have different values for leading and trailing piles, depending on pile-to-pile spacing. It is acceptable to use a weighted average pile lateral capacity when comparing it with the maximum lateral demands on the piles. This is a reasonable estimate, since the pile cap (basemat) is stiff enough to redistribute the forces between individual piles.

The load combinations used for pile capacity checks are the service load combinations, since the typical capacities provided by the geotechnical engineer are allowable capacities.

8.13 SOIL-BEARING PRESSURE CHECK

For TG foundations, the supporting soil typically has a bearing capacity of 4,000psf or higher in order to control foundation settlements. The actual design bearing capacity shall be determined by geotechnical engineers.

Similar to the pile capacity check, the load combinations used for the soil-bearing capacity check are service load combinations. A simple hand calculation or FE analysis can be used to determine the maximum bearing pressure under normal operational loads, emergency operation or catastrophic equipment loads, and seismic loads. For emergency operation or catastrophic equipment and seismic loading scenarios, the soil-bearing capacity can be increased by 33%, upon approval by the geotechnical engineer.

Because of the massive nature of the TG foundation, there is typically no net tension (i.e., uplifting) underneath the basemat. However, for very high seismic zones, uplift needs to be considered. An FE nonlinear analysis may become necessary to determine the maximum soil pressure in such cases.

8.14 STABILITY EVALUATION

TG foundations have large basemat footprints and are very heavy compared to other power plant concrete foundations. Therefore, sliding, overturning, and buoyancy are usually not an issue. However, for TG foundations located in high seismic zones, the foundations may slide because large horizontal seismic forces. The design engineer can estimate the sliding using the energy balance method (ASCE 4, 1998) to ensure sufficient gap/separation is provided between the TG foundations and other surrounding structures. Note that the seismic lateral force of basemat is typically not required to be included in the stability check for soil-supported foundations.

REFERENCES

ACI (American Concrete Institute). (2014). "Building code requirements for reinforced concrete." *ACI 318-14*, Detroit.

ACI. (1995). "Effect of restraint, volume change, and reinforcement on cracking of mass concrete." *ACI 207.2R-95*, Detroit.

ACI. (2006). "Code requirements for environmental engineering concrete structures." *ACI 350-06*, Detroit.

ASCE. (2010). "Minimum design loads for buildings and other structures." *ASCE/SEI 7-10*, Reston, VA.

ASCE. (1998). "Seismic analysis of safety-related nuclear structures and commentary." *ASCE 4-98*, Reston, VA.

ASCE Task Committee on Seismic Evaluation and Design of Petrochemical Facilities. (2011). *Guideline for seismic evaluation and design of petrochemical facilities*, 2nd Ed. Reston, VA.

British Standards Institution. (1974). "Code of practice for foundations for machinery: foundations for reciprocating machines." *CP2012-1*, London. (old British practice, withdrawn)

DIN (Deuches Institut für Normung). (1988). "Machine foundations, flexible structures that support machines with rotating elements." *DIN 4024-1*, Berlin. (old German practice, withdrawn)

Reineck, K. H., and Novak, L. C., eds. (2010). *Further examples for the design of structural concrete with strut-and-tie models*. SP-273, American Concrete Institute, Farmington Hills, MI, 288.

CHAPTER 9
EMBEDDED ITEMS

9.1 INTRODUCTION

To properly install and secure the turbine generator equipment and its auxiliaries to the foundation, embedded items, such as anchor bolts, jacking posts, embedded plates, and shear keys, are required. This chapter addresses the function, arrangement, installation, and design of various types of embedded items in turbine generator foundations.

This chapter highlights a number of design issues related to the design of large diameter anchor bolts in TG foundations. For more general information on anchorage, see ASCE (2012), *Anchorage Design for Petrochemical Facilities*.

9.2 TYPES OF EMBEDDED ITEMS IN TURBINE GENERATOR FOUNDATIONS

9.2.1 Anchor Bolts and Sleeves

Anchor bolts used in TG foundations are partially threaded rods, with heavy nuts at the bottom, either cast-in-place or through-bolts, or grouted in pockets. Hooked bolts (L-bolt and J-bolt) should not be used in TG foundations (ACI 349), as they do not provide satisfactory performance in securing dynamic equipment.

Anchor bolt sleeves can be provided to allow for pretensioning and small adjustments in the field as required. Partial sleeves are made of polyethylene, either smooth or corrugated, while full length sleeves are steel pipes. The anchor sleeve diameter should be at least 1.0 in. (25 mm) larger

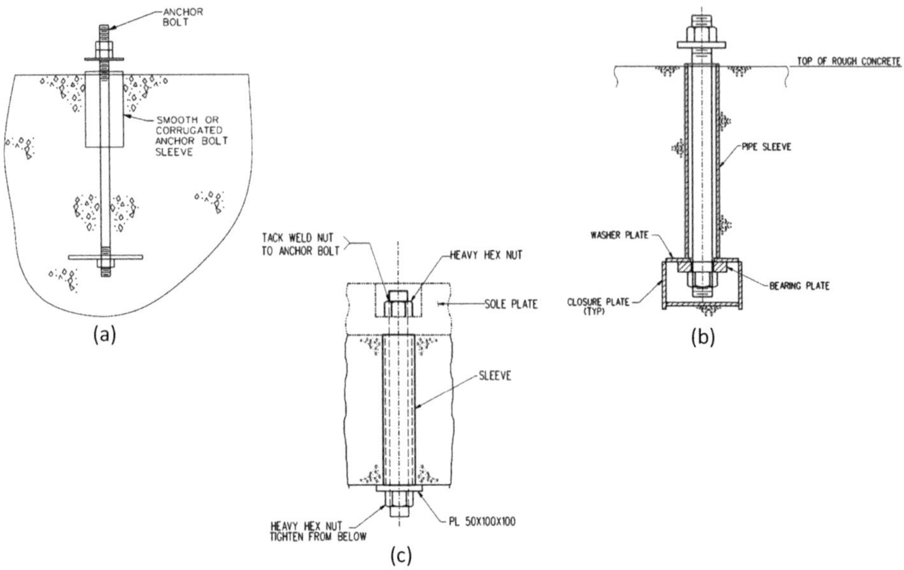

Fig. 9-1. Anchor with partial sleeve (a), anchor bolt with full-length sleeve (b), and anchor bolt with full-length sleeve for through-bolt (c)

than the anchor bolt nominal diameter. During concrete placement, the sleeves should be shielded or plugged to prevent the entry of concrete. Fig. 9-1 depicts examples of cast-in-place anchor bolts and sleeves.

Anchor bolt sleeves provide three advantages:

1. They allow for minor repositioning of anchor bolts to match the installed equipment position. Note that bolts larger than 1.5 in. in diameter become too stiff to move even with a sleeve and cannot be repositioned.
2. For anchor bolts with a significant preload, sleeves will prevent bonding to the concrete and allow for free stretching or elongation of the anchor bolts.
3. Sleeves for through-bolts facilitate the tightening of bolts from underneath.

Anchor bolts that are required to have a significant preload in the final installed condition should be installed in sleeves, or the shank of the anchor bolt must be coated or covered with a bond breaker to allow for bolt elongation after pretensioning.

While the information on anchor bolt type, size, number, location, elevation, and preload is provided by the TG manufacturers, it is usually the TG foundation design engineer's responsibility to determine the overall length of the anchor bolts.

Fully grouted anchor bolts were sometimes proposed by overseas machine vendors. Normally, the process requires casting a pocket in the foundation first, then dropping the bolt and filling out the pocket with grout. However, determining the bolt capacities for this scenario may not be found in building codes. Some practitioners choose to treat the grout as concrete when doing so. Sometimes building officials may request testing to justify such an approach.

9.2.2 Embedded Plates and Shear Keys at TG Equipment Foundation Interfaces

Besides anchor bolts, which are primarily for transferring tensile forces from the TG equipment to the foundation, other important structural elements are embedded plates and shear keys that primarily transfer shear forces from the equipment to the foundation.

Embedded plates, with shear keys and/or headed bolts, can be used as the foundation connections for the TG axial stop blocks, anchor support plates, exhaust hood anchors, and centerline guides. Fig. 9-2 shows how the equipment base is connected to the embedment plate through an anchor pin and a centerline guide (or an axial stop block).

Shear keys (or lugs) in TG foundations are steel elements such as heavy steel plates (2 in. to 4 in. thick), W-shapes, or tubes that are welded to the underside of the equipment support embedded plates, and either cast in place with concrete or grouted in pockets. They provide a shear loading transfer path through direct bearing with concrete or grout, transferring large shear forces, such as those resulting from thermal expansion and accidental scenarios, from the TG equipment into the concrete foundation.

Plate shear keys underneath an embedment plate can be arranged to transfer shear forces from two directions. The same plate can also have welded studs or bolts that transfer axial forces into the foundation. Fig. 9-3 provides example details of an embedment plate with shear keys and headed bolts.

The TG manufacturer may provide a suggested shear key design and layout, but it is the design engineer's responsibility to ensure that the shear keys and the surrounding concrete are structurally adequate to transfer the shear loading into the foundation.

9.2.3 Horizontal Adjustable Device—Jacking Posts

Jacking posts are temporary (or reusable) devices made of rolled steel shapes or built-up sections for horizontal alignment of TG equipment. The TG equipment position may be adjusted by placing a jack against the jacking posts. Jacking posts are usually located alongside the equipment

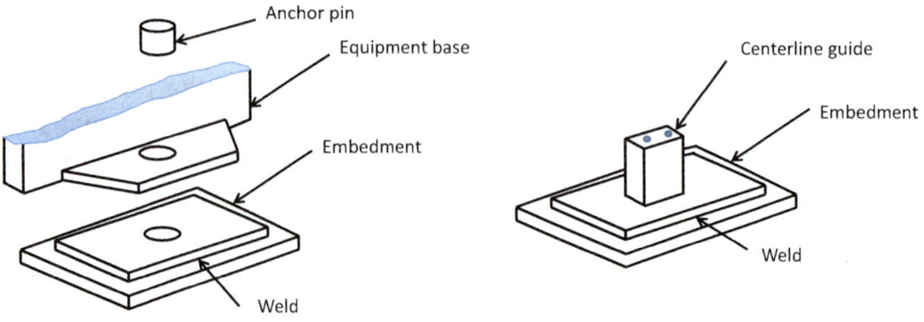

Fig. 9-2. Embedment plate connection to equipment base

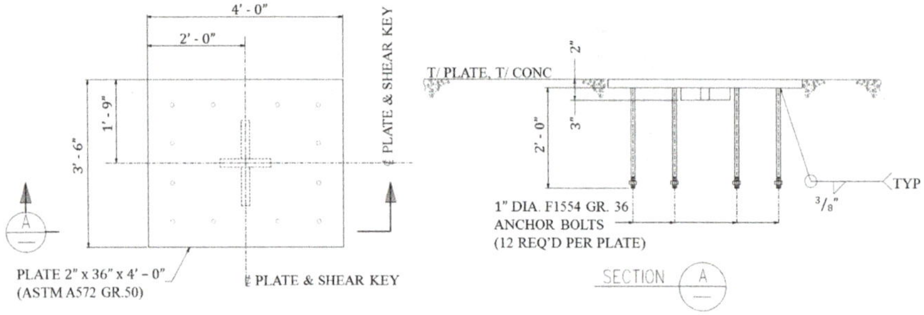

Fig. 9-3. Example embedded plates with shear keys and headed bolts

for transverse adjustments perpendicular to the turbine generator shaft centerline. Jacking posts may also be used to aid in axial adjustments, along the rotor. Fig. 9-4 depicts a typical layout of jacking posts on a CTG foundation.

Jacking posts can be either cast-in-place with concrete, and flame-cut to be flush with the top of concrete after use, or be placed in a sleeve or anchored to the side of the foundation, thus becoming reusable. The location, projection over the top of the foundation, and number of jacking posts required are provided by the TG manufacturer.

Fig. 9-5(a) illustrates a wide flange jacking post that is partially embedded in concrete with its top 12 in. (305 mm) projecting over the top of concrete. After the installation alignment is complete, the jacking post will be cut flush with concrete. Fig. 9-5(b) provides an example of a jacking screw that may be attached to the top of a jacking post. The jacking screw is only suitable for equipment with a gross weight less than 200 kips (890 kN).

For large jacking forces (>100 kips), additional reinforcement may be required to help transfer the load into the foundation. Fig. 9-6 illustrates

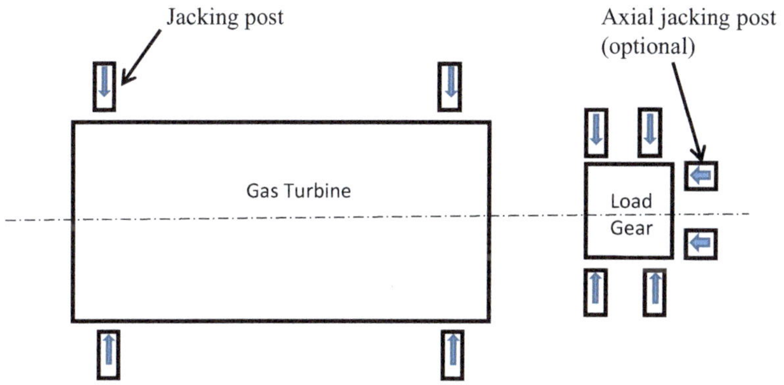

Fig. 9-4. Jacking posts on a CTG foundation

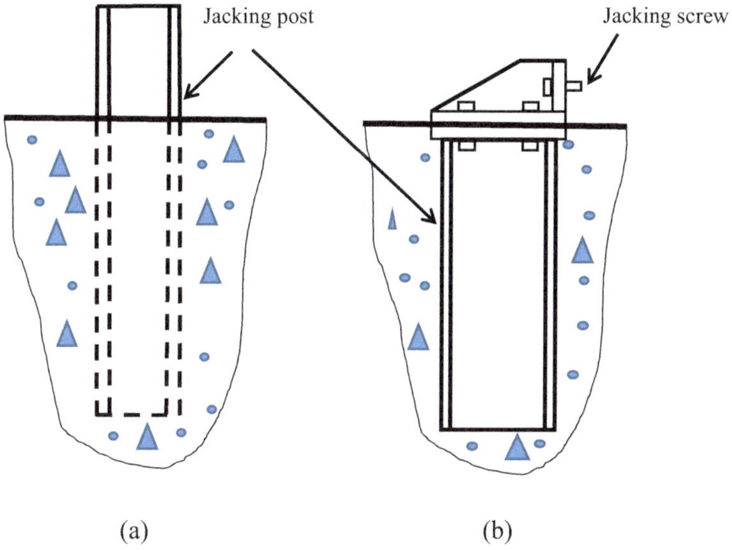

Fig. 9-5. Jacking post cast in concrete

additional horizontal weldable steel reinforcement used for transferring large jacking post forces.

Jacking posts may also be designed as a removable attachment to the foundation. Fig. 9-7 shows a jacking post attached to the side of a TG foundation. Fig. 9-8 shows a jacking post that can be placed in a sleeve embedded in concrete. The jacking posts in both configurations are reusable.

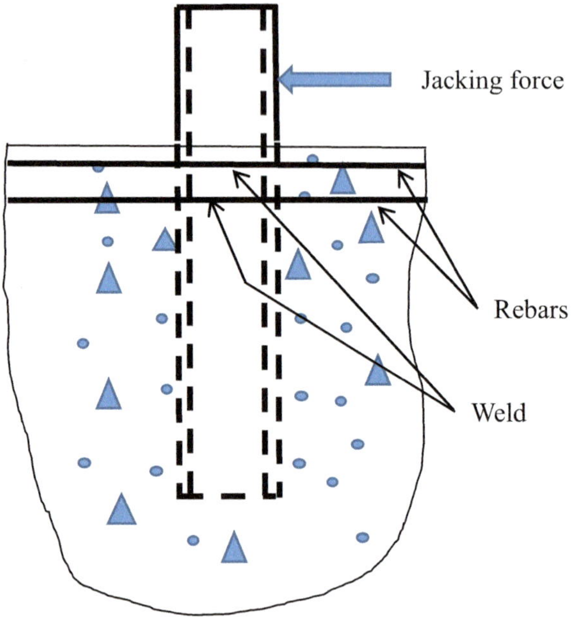

Fig. 9-6. Additional reinforcement at embedded jacking post

The TG manufacturer may provide a suggested jacking post design and layout, but it is the design engineer's responsibility to ensure that the jacking posts and the surrounding concrete are structurally adequate to transfer the jacking forces during the equipment installation or maintenance into the TG foundation.

The interference between jacking posts and TG foundation reinforcement should be carefully accounted for in the design. Sometimes predrilled holes in the jacking posts may be necessary to allow reinforcing rebars to pass through rather than terminating rebars at the jacking posts.

9.2.4 Special Embedded Items

Some TG manufacturers require customized embedded items that are designed to accommodate large bolt pretension, or special connections between the TG equipment bolts and sole plates and the foundation.

For example, the embedded anchor box shown in Fig. 9-9 is designed to be able to resist large bolt pretension. Large diameter high strength (1¾ in. diameter, ASTM A615, 75 ksi) threaded rods are used to meet the pre-

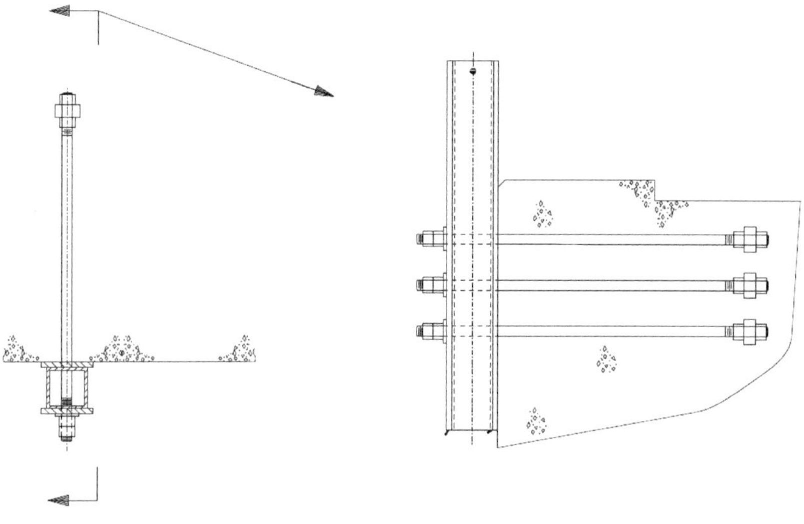

Fig. 9-7. Jacking post at side of foundation

tension requirements. Special ball nuts and conical washers are in the box to help distribute the pretension into the concrete. Close coordination between the TG foundation designer and the TG manufacturer is essential for successful design and fabrication of these types of embedment.

9.2.5 Miscellaneous Embedded Plates

Besides supporting the turbine and generator through anchor bolts and embedded plates with shear keys, the TG foundation also provides support and restraint for miscellaneous pipes, ductworks, electrical cable trays, and conduits routed within and around the perimeter of the TG foundation. These miscellaneous components are typically supported by embedded plates, cast-in-place inserts, and post-installed anchors. For post-installed anchors, both mechanical and adhesive anchors can be used.

Figs. 9-10 through 9-13 show samples of miscellaneous embedded plates in a TG foundation. Locations of these embedded plates should be clearly identified and addressed in the structural calculations and properly detailed to maintain their structural integrity. Reinforcing around these elements should be planned and considered in the development of construction drawings.

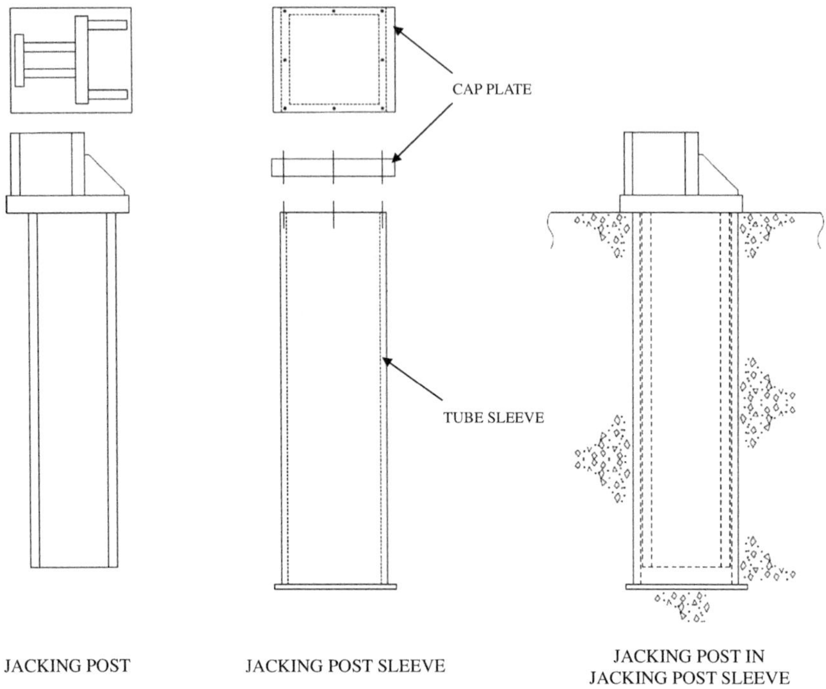

Fig. 9-8. Example jacking post sleeve embedded in concrete

9.3 STRUCTURAL DESIGN FOR EMBEDDED ITEMS

9.3.1 Cast-in-Place Anchor Bolt Design Considerations

In a TG foundation, anchor bolts are often designed to resist tensile loading only, while the shear loading is transferred to the TG foundation through embedded plates and shear keys. This design concept simplifies the anchorage design.

The general design philosophy in concrete anchorage design is to have a ductile design, or to design the anchorage as a ductile connection to the concrete. A ductile connection is controlled by the yielding of steel elements, including the anchor and reinforcement, with large deflection, redistribution of loads, and absorption of energy, rather than a sudden loss of capacity of anchorage resulting from a brittle failure of the concrete. ACI 318 provides the requirements for ductile anchorage designs. Some types of nonbuilding structures may use the foundation anchor bolt as a yield mechanism to provide structural ductility. For example, ASCE 7,

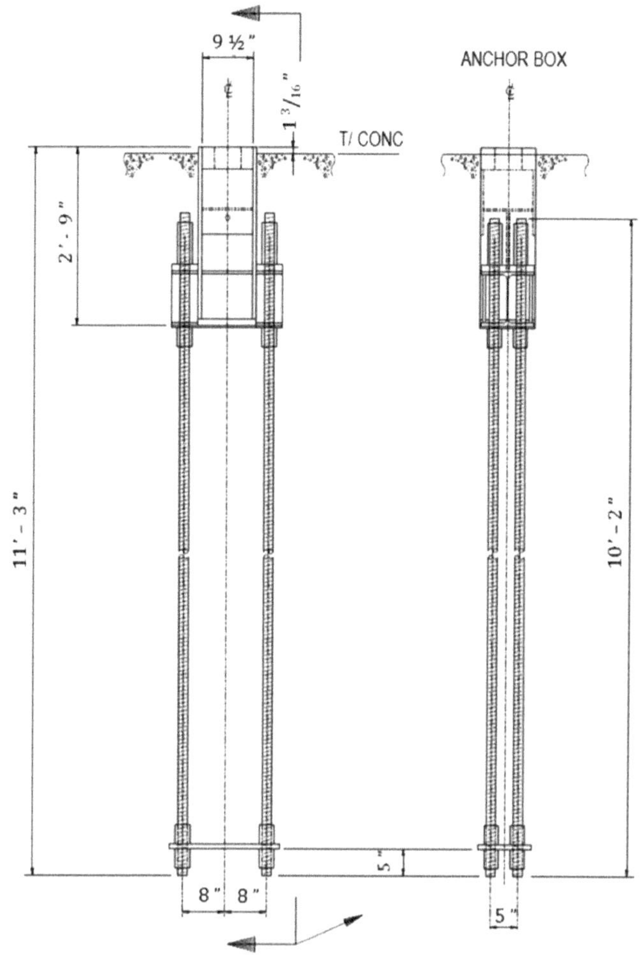

Fig. 9-9. Example special embedded anchor box

Section 15.7.5, requires tanks and vessels to be intentionally designed to create a plastic mechanism of tensile yielding in the anchors.

Most anchor bolts in TG foundations are large diameter bolts (1.5 in. to 4 in.). When bolt preloading is specified by the TG manufacturer, it is very likely that the bolts are oversized for the actual design load combinations. To design a concrete anchorage for oversized bolts and to have a concrete anchorage capacity not less than the tensile strength of the bolts is extremely difficult, if not impossible, to achieve in a TG foundation.

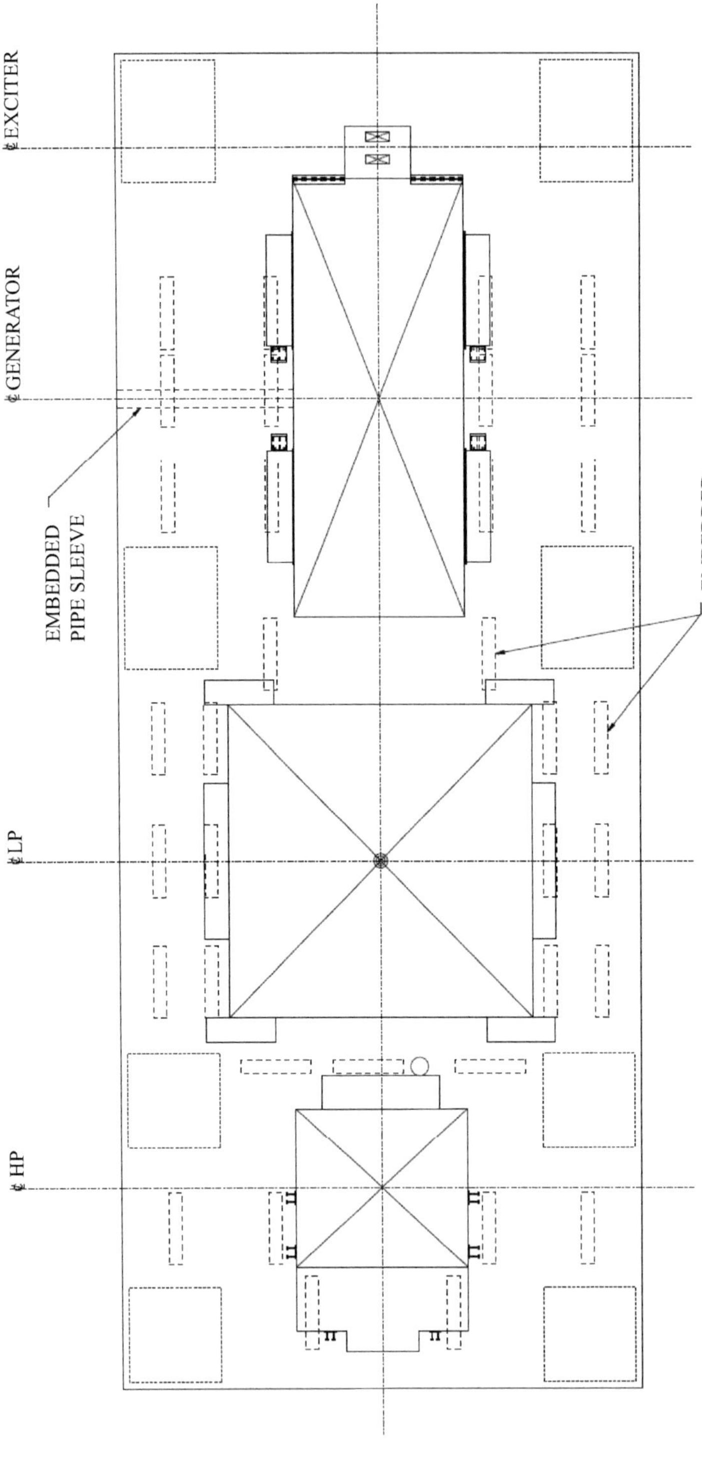

Fig. 9-10. Plan view: miscellaneous embedded plates at underside of an STG tabletop

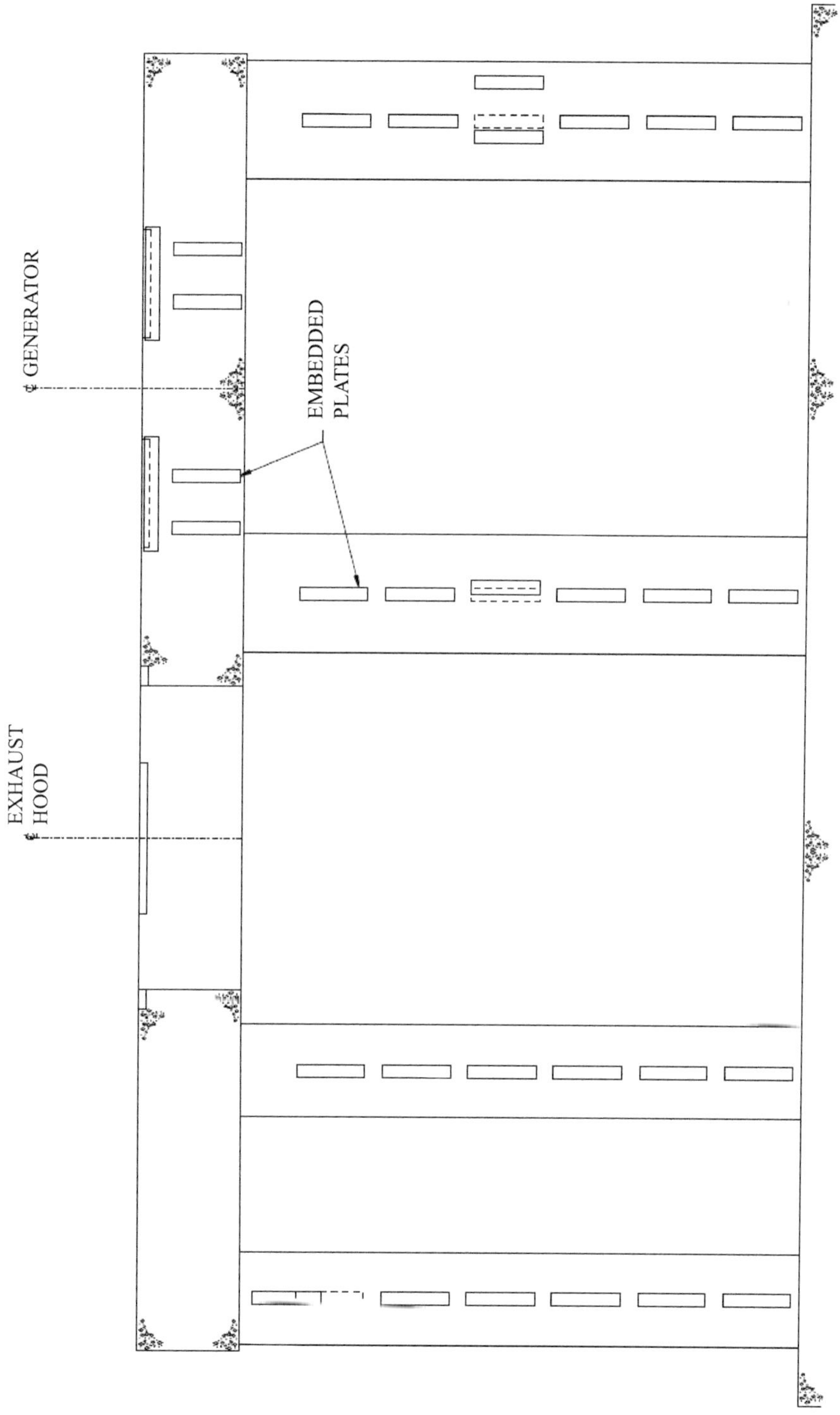

Fig. 9-11. Elevation view: miscellaneous embedded plates on columns and side faces of an STG foundation

Fig. 9-12. Pipes attached to embedded plates on column
Source: Courtesy of Kiewit Engineering Group; reproduced with permission.

Fig. 9-13. Attachments at side of tabletop
Source: Courtesy of Kiewit Engineering Group; reproduced with permission.

On the subject of ductile anchorage design, the ASCE Task Committee on Anchorage Design (2012), Section 3.4 states, "If the anchor reinforcement is properly designed and developed to prevent concrete failure, the resulting connection can be considered as ductile. This philosophy is consistent with reinforced concrete design principles." This Committee shares the same opinion for anchorage design for TG foundations. Proper anchor reinforcement design means appropriately applying the overstrength factor (Ω_0) to the design loading, and checking the concrete for breakout, anchor pullout, and side-face failure modes. If the concrete itself cannot provide sufficient strength, reinforcement must be provided and properly detailed around the anchor bolts to prevent nonductile concrete failure.

The ductile yield mechanism may be achieved by an additional anchor reinforcement for concrete breakout, either shear or tension or both.

Seismic design per ACI 318 includes subsections 17.2.3.1 through 17.2.3.7 and provides four options for determining the required anchor or attachment strength to protect against non-ductile tensile failures. Option (d) states, "The anchor or group of anchors shall be designed for the maximum tension obtained from design load combinations that include E, with E increased by Ω_0. The anchor design tensile strength shall satisfy the tensile strength requirements of 17.2.3.4.4." For this option, the seismic loads are increased by the Ω_0 factor and the anchor strength is reduced from their nominal tensile strengths by an additional reduction factor of 0.75. This option can be used for the large anchor bolt anchorage design in TG foundations.

9.3.1.1 Anchor Bolt Reinforcement Design. Steel reinforcement may be used as anchorage reinforcement to cross a potential breakout concrete cone to provide the required anchorage capacity. Anchor reinforcement can be designed as supplementary reinforcement or to transfer the full design loads into the concrete as follows:

1. Supplementary reinforcement acts to restrain the potential concrete breakout but is not designed to transfer the full design load from the anchors into the structural member. ACI 318 requires the use of a reduction factor φ for concrete failure modes except for pullout and pryout strengths.
2. Anchor reinforcement is used to transfer the full design load from anchors into the structural member, when the factored tensile loading demand exceeds the concrete breakout strength. Anchor reinforcement is required per ACI 318, Chapter 17, for anchors with diameter greater than 4 in. for concrete shear breakout, or anchors with embedment length larger than 25 in. for concrete tensile breakout.
3. Anchor reinforcement has to be designed to fully develop its strength on both sides of the cracking planes it passes through.

Anchor reinforcement may also be designed using the strut-and-tie concept per ACI 318, Chapter 23.

9.3.1.2 Tensile Failure Cases for Anchor Bolt Design in TG Foundations. Various types of steel and concrete failure modes for anchors are shown in

ACI 318, Fig. R.17.3.1. Specifically, the following three types of tensile failure cases should be considered in a TG foundation.

1. *Concrete Breakout*: Concrete breakout, under either tensile or shear loading, occurs when a volume of concrete surrounding the anchor or group of anchors separates from the foundation or structural element. The potential separating surface is expected to occur at a 35-degree angle from the horizontal plane for tension or the vertical plane for shear. The capacity of concrete breakout can be calculated per ACI 318, Chapter 17. When the tensile loading exceeds the concrete breakout capacity, anchor reinforcement should be provided and developed for the applicable tensile loading.
2. *Side-face Blowout*: Anchors with minimal edge distance and deep embedment or thin side cover may be subject to a side-face blowout failure, where the concrete face around the anchor head spalls while the concrete near the top of anchor experiences no breakout. Side-face blowout strength may be calculated per ACI 318, Section 17.4.4. Anchor reinforcement should be provided when the concrete side-face blowout strength capacity is exceeded.

   Side-face blowout for a single anchor is calculated per ACI 318, Eqs. 17.4.4.1 and 17.4.4.2. If ties or spirals are used to prevent side-face blowout, the rebar should be placed close to the anchor. Spirals are better than ties for resisting side-face blowout. Research shows that "transverse reinforcement (ties) do not increase the side-face blowout capacity and a larger amount of transverse reinforcement installed near the anchor head only increases the magnitude of load that is maintained after the side-face blowout failure has occurred (that is, ductility is increased)" (ACI 318, Section 3.5.2, f). Therefore, the TG design engineer should consider increasing concrete edge distance and/or using a large bearing plate before adding supplemental reinforcement to prevent side-face blowout.
3. *Anchor Pullout*: Per ACI 318, pullout failure corresponds to the anchor device or a major component of the device sliding out from the concrete without breaking out a substantial portion of surrounding concrete. Based on ACI 318, Eq. 17.4.3.1, pullout strength depends on the size of the headed stud or bolt, and strength of concrete. Anchor reinforcement may not improve the pullout capacity in this case.

9.3.1.3 Preloading. Preloading, or clamping force, is developed by tensioning the anchor. Sufficient clamping force can maintain the critical alignment of the machine. The clamping force is intended to allow smooth transmission of unbalanced machine forces into the foundation such that the machine and foundation can act as an integrated system. The pretensioned anchors provide both moment and axial capacity at the interface. In addition, a calculation can be done to determine how much pre-compressed stress is required at the edge of baseplate from tensioning anchor bolts to balance an unbalanced vibrated machine force. However, to simplify the design, ACI 351.3R suggests a minimum anchor clamping force of 15% of the anchor material yield strength when specific values are not provided by TG machine manufacturer.

Anchor bolt preload can be used to provide horizontal capacity of the anchor system as given by the following equation:

$$T_{min} = \frac{F_r}{\mu} - W_a \tag{9-1}$$

where

T_{min} = minimum required anchor tension;
F_r = maximum horizontal dynamic force;
μ = coefficient of friction;
W_a = equipment weight at anchorage location.

This equation ensures that the tie-down force is large enough to resist the maximum horizontal dynamic force after friction is considered. (Note that the strength reduction factor should be $\varphi = 0.75$.)

9.3.2 Post-Installed Grouted Anchor Rods

Anchor rods for attaching turbines and generators to their foundations are sometimes post-installed by installing a headed anchor in an oversized pocket in the foundation and grouting the anchor in place with cementitious grout. Although the design of these anchors has yet to be codified in the United States, references are available for their design (Miltenberger 2001; Subramanian and Cook 2004). The design approach for these anchors is similar to a combination of the concrete capacity design (CCD) method (Fuchs et al. 1995) and the Uniform Bond Stress model (Cook et al. 1998). The CCD method is the basis of provisions for cast-in-place headed anchors within ACI 318, and the uniform bond stress model combined with the CCD method is the basis for design provisions for adhesive anchors in ACI 318.

9.3.3 Jacking Post Design

For jacking post design, the loading can be estimated as the machine weight at installation multiplied by the coefficient of friction between the jacking post and the machine (steel on steel). For a jacking post embedded in the TG foundation (Figs. 9-5 and 9-6), a conservative design approach is to use PCI, Eq. (6-75), for steel haunches. The nominal strength of the jacking post is as follows.

$$V_c = \frac{0.85 f'_c b L_e}{1 + \frac{3.6e}{L_e}} \tag{9-2}$$

where

V_c = nominal strength of the section controlled by concrete;
L_e = embedment length;
b = effective width of compression, which is generally the flange width;
f'_c = concrete compressive strength;
$e = a + L_e/2$ (eccentricity of force to half embedment length).

The equation assumes that the jacking post is supported by two compression regions of concrete (Marcakis and Mitchell 1980). See Fig. 9-14.

Additional reinforcement, as shown in Fig. 9-6, may be checked per PCI, Eq. (6-76):

$$V_r = \frac{2A_s f_y}{1 + \frac{6e/L_e}{4.8s/L_e - 1}} \tag{9-3}$$

where

V_r = nominal shear strength provided by reinforcement;
f_y = yield strength of reinforcement;
s = spacing of reinforcement;
A_s = cross area of reinforcement.

Then $V_n = V_c + V_r$ and $V_u \le \varphi V_n$, where strength reduction factor $\varphi = 0.85$.

Alan H. Mattock and G. H. Gaafar (1982) developed different equations as follows:

$$V_n = 0.85 f'_c \beta_1 b L_e \left(\frac{0.58 - 0.22\beta_1}{0.88 + a/L_e} \right). \tag{9-4}$$

The maximum moment of post is given by

$$M_{nmax} = V_n \left(a + \frac{V_n}{1.7 f'_c b} \right). \tag{9-5}$$

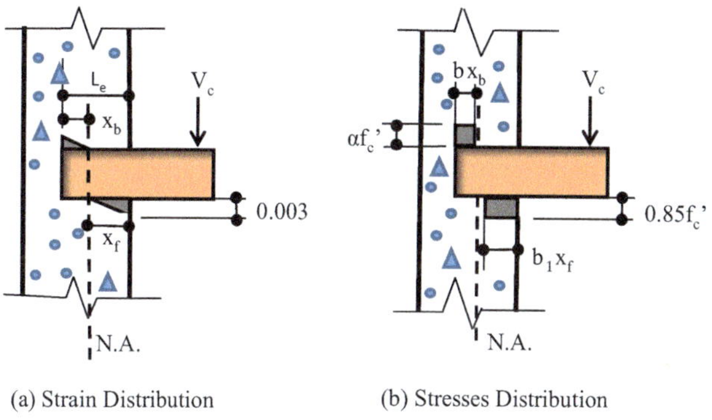

Fig. 9-14. Equations and assumptions of steel haunches

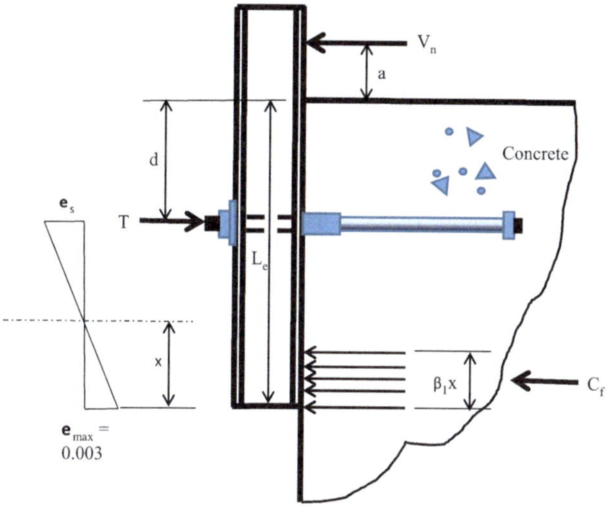

Fig. 9-15. Side jacking post design

where β_1 is a factor relating depth of equivalent rectangular compressive stress block to neutral axis depth as specified in Section 10.2.7 of ACI 318.

Jacking posts attached to the side of the foundation (Fig. 9-15) can be analyzed as a cantilever beam. The end of the jacking post bears against the concrete, while the anchor bolts provide tension resistance for the jacking load. Using ACI 318 reinforced concrete assumptions for bending and compression, the following equations may be used to determine the nominal strength of the jacking post:

Strain can be calculated from the maximum concrete strain and plane-remains-plane assumption as

$$\varepsilon_s = 0.003\frac{L_e - x - d}{x} \tag{9-6}$$

1. The tension force in the anchor is

$$T = E_s\, \varepsilon_s\, A_s \tag{9-7}$$

2. The compression force on the concrete surface is

$$C_f = 0.85 f_c' b \beta_1 x \tag{9-8}$$

3. The equilibrium equations are

$$T = V_n + C_f \tag{9-9}$$

$$M_{max} = V_n(a+d) = C_f(L_e - d - \frac{1}{2}\beta_1 x). \tag{9-10}$$

where

C_f = total concrete force of equivalent rectangular stress block;
d = distance from center of anchors to concrete surface;
x = concrete strain region, to be solved from these equations;
A_s = total cross area of anchor bolts.

The "x" value can then be solved for from the five equations (Eqs. 9-6 to 9-10). Then the nominal strength of the jacking post can be calculated. The anchor bolt size, embedment, and additional reinforcement (if required) should be designed to resist the tensile force T (Eq. 9-7) per ACI 318, Chapter 17.

9.3.4 Shear Key Design

Frictional resistance and shear capacity of anchors typically are considered a part of the shear transfer mechanism. However, for turbine generator foundations, shear lugs provide a more effective mechanism to transfer large shear loads and satisfy ductile anchorage design requirements. Shear lugs can be assumed to resist all shear loadings. The shear lugs may be designed per ACI 349 or AISC *Steel Design Guide for Base Plates and Anchor Bolts Design*. The failure shear planes initiated from a shear key as per ACI 349 are assumed to have a 45-degree spreading angle from the horizontal plane, different from the 35-degree angle assumed in ACI 318. The portion of the key bearing against the grout over the top of concrete should not be included in the design calculation. Although the actual bearing load against the shear key is probably higher near the top of concrete and reduce toward the bottom of the lug, the AISC *Steel Design Guide for Base Plates and Anchor Bolts Design* assumes the bearing is uniform for the purpose of design.

Where multiple shear keys are used, there is an additional design requirement. The shear strength of shear lugs by bearing should not exceed the shear strength between shear keys as defined by a shear plane between the shear lugs, shown in RD.11.1 of ACI 349.

Additional reinforcement, such as hairpins and tiebacks, for shear breakout per ACI 318, Chapter 17.5, may be used in combination with shear lugs to allow shear loads to be adequately transferred from the attachment to the grout and then to the concrete TG foundation.

9.3.5 Embedded Plate Design

Miscellaneous embedded plates are designed with headed studs, anchor bolts, or shear lugs to transfer loads to the concrete TG foundation.

Some embedded plates may be prescribed by the TG machine manufacturer, while others are required by mechanical/electrical engineers who need supports off the turbine generator foundation for their commodities. Close coordination between all disciplines is important to avoid clash and rebar congestion during construction.

Meeting the ductile design requirements for embedded plates may be difficult given the short, closely spaced headed stud anchors. The attachment should be designed to yield first, or the anchor design strength reduced to minimize the possibility of a brittle failure per ACI 318. Alternatively, longer welded rebar may be used to replace welded studs to avoid brittle failure of the concrete. Similar to anchor bolt design, use of strength reduction or overstrength factor, Ω_0, may be more practical.

9.4 GROUT POCKETS, GROUT DAMS, AND GROUT

9.4.1 Grout Pockets

Grout pockets (Fig. 9-16) are recesses in TG foundations that are reserved for future grouting after the machine is aligned and finally positioned in place. Grout pockets are required for shear keys or grout pins, Fixators, and adjusting bolts. The TG manufacturer usually depicts the required pocket locations and sizes on their foundation requirement drawings. When sizing or modifying the grout pockets, if required, the following should be accounted for:

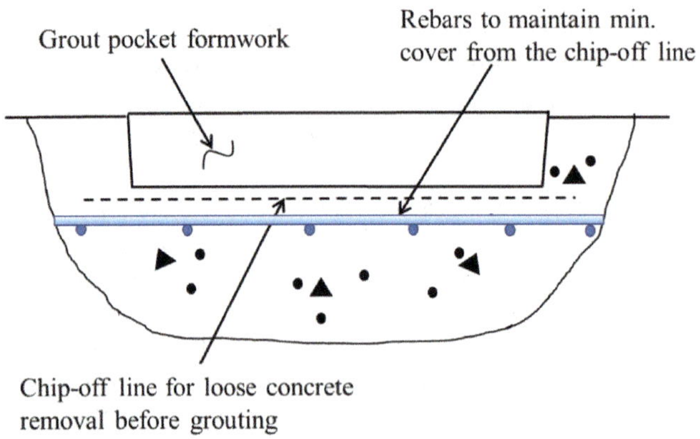

Fig. 9-16. Typical grout pocket

1. Grout pockets generally have a depth varying from 1.0 in. (25 mm) to 4.0 in. (102 mm) unless specified to accommodate grout pads or adjustable support devices, such as Fixators (see Figs. 11-8 to 11-11).
2. Installation methods may require additional depth to be added to the grout pocket.
3. The grout pocket must be sufficiently deep to accommodate shear keys or grout pins.
4. The concrete quality at the bottom of the grout pocket may be impaired by an accumulation of water and fine aggregate. Therefore, it is recommended to chip off ½ in. (12 mm) concrete at the pocket bottom to ensure sound bonding between grout and concrete (see Fig. 9-16).
5. The grout pocket should be sized to provide a minimum clearance of 2.0 in. (50 mm) between the side of the grout pocket and the edge of the TG foundation baseplates. Additional clearance should be provided on the side where the grout placement will be made. If adjustable support devices are used, additional clearance will be needed where wrench access is required.
6. The grout pocket may need to be contoured to accommodate adjacent hardware such as jacking posts and conduit stub-ups.

9.4.2 Grout Dams

Grout dams are light steel hardware that are used along a free concrete edge in order to form a grout pocket when the pocket is close to the edge of the TG foundation (see Fig. 9-17). Grout dams help contain the grout during grout placement. When the grout dam is in close proximity to

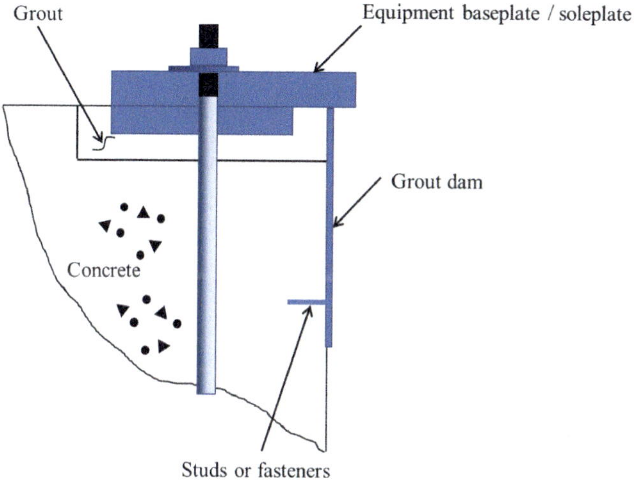

Fig. 9-17. Grout dam at a concrete free edge

equipment, the dams should be affixed such that there are no projections beyond the face of the grout dam.

9.4.3 Grout

The grout is a construction material to be placed between equipment sole plates and the concrete TG foundation. The selection of a grout type is determined by the design engineer or the turbine generator manufacturer.

Cementitious grouts can have compressive strengths as high as 15,000 psi (48.5 MPa) and cost less than epoxy grouts. Flowable nonshrink cementitious grout has been widely used for turbine generator foundations for its successful service history, wide availability, temperature capability, and workability.

Epoxy grouts have better vibration resistance and chemical resistance to process fluids and oils than cementitious grouts. In addition, they are formulated with better physical properties than cementitious grouts, offering higher compressive, tensile, and flexural strengths, as well as increased adhesion and shear capacity. However, epoxy grouts have temperature limitations on working temperature, and since they have different thermal expansion coefficients from concrete, creep and shrinkage cracking may be more likely to develop than for cementitious grouts. Due to such temperature considerations and limitations, nonshrink cementitious grout, complying with ASTM C 1107 requirements, are still recommended by many TG manufacturers.

Turbine dynamic loads constantly produce horizontal, vertical uplifting, and downward forces on the sole plates and the grout. During operation, a steam turbine foundation can be subjected to high ambient temperatures generated by the machine components. Epoxy grouts can perform at base plate temperatures under 130°F (55°C), whereas cementitious grout can operate at temperatures up to 400°F (200°C). When specifying a grout for a specific dynamic loading system, severity of the shock loads, temperature requirements, ease of placements and other considerations should be closely examined. Grout selection should also consider tests, such as the performance and flow evaluation test, specified in ACI 351.1R, or prior experience with the grout in similar situations. Once a grout type is selected, the mixing, placement, and curing should be per the manufacturer's instructions.

The grout should fully occupy the space between the sole plates and the concrete TG foundation. Large sole plates or skids may cause problems as the grout may not fully flow into all the spaces. Grout air holes in sole plates can be used to help the grout fully occupy the space beneath the plates.

ACI 351.1R, "Report on Grouting between Foundations and Bases for Support of Equipment and Machinery," provides additional information on grouting.

REFERENCES

ACI (American Concrete Institute). (2014). "Building code requirements for structural concrete and commentary." *ACI 318-14*, Farmington Hills, MI.

ACI. (2013). "Code requirements for nuclear safety-related concrete structures (ACI 349-13) and Commentary." *ACI 349-13*, Farmington Hills, MI.

ACI. (2012). "Report on grouting between foundations and bases for support of equipment and machinery." Reported by ACI Committee 351, *ACI 351.1R-12*, Farmington Hills, MI.

ACI. (2004. Reapproved 2011). "Foundations for dynamic equipment." Reported by ACI Committee 351, *ACI 351.3R-04*, Farmington Hills, MI.

AISC (American Institute of Steel Construction). (2006). *Steel design guide 1*, 2nd Eds., J. M. Fisher and L. A. Kloiber, *Base Plate and Anchor Bolt Design*. American Institute of Steel Construction, Chicago.

AISC. (2005). "Seismic provisions for structural steel buildings." *AISC 341-05*, Chicago.

ASCE. (2010). "Minimum design loads for buildings and other structures." *ASCE/SEI 7-10*, Reston, VA.

ASCE Task Committee on Anchorage Design. (2012). *Anchorage design for petrochemical facilities*. Reston, VA.

ASCE Task Committee on Turbine Foundations of the Fossil Power Committee and the Nuclear Energy Committee of Energy Division. (1987). *Design of large steam turbine-generator foundations*. Reston, VA.

ASTM International. (2008). "Standard specification for pipe, steel, black and hot-dipped, zinc-coated, welded and seamless." A36/A36M-08. ASTM International, West Conshohocken, PA.

ASTM. (2010). "Standard specification for alloy-steel and stainless steel bolting for high temperature or high pressure service and other special purpose applications." A193/A193M-10a. ASTM International, West Conshohocken, PA.

ASTM. (2010). "Standard specification for carbon steel bolts and studs, 60,000 PSI tensile strength." A307-10. ASTM International, West Conshohocken, PA.

ASTM. (2007). "Standard specification for washers, steel, plain (flat), unhardened for general use." F1554-07a. ASTM International, West Conshohocken, PA.

California Department of Transportation. (2012). *Bridge design aids, Section 5-81, Anchorage to concrete*. Sacramento, CA.

Cook, R. A., Kunz, J., Fuchs, W., and Konz, R. C. (1998). "Behavior and design of single adhesive anchors under tensile load in uncracked concrete." *ACI Struct. J.* 95(1), (Jan-Feb).

Fuchs, W., Eligenhausen, R., and Breen, J. (1995). "Concrete capacity design (CCD) approach for fastening to concrete." *ACI Struct. J.* 92(1), (Jan-Feb), 73-93. Related discussions 92(6), (Nov-Dec), 787–802, 791–792, 798–799.

General Electric. (2015). *Turbine generator foundations. GEK 63383e.*

Marcakis, K., and Mitchell, D. (1980). "Precast concrete connections with embedded steel members." *PCI J.*, 25(4), 88–116, July-August.

Mattock, A. H., and Gaafar, G. H. (1982). "Strength of embedded steel sections as brackets." *ACI J.*, (Mar-Apr).

Miltenberger, M. (2001). *Capacity design of grouted anchors*, Transactions, Paper 1790, SMiRT 16, Washington, DC, August.

PCI (Precast/Prestressed Concrete Institute). (2010). *PCI design handbook, MNL-120-10*, 7th Ed., Chicago.

Subramanian, N., and Cook, R. A. (2004). "Behaviour of grouted anchors." *Indian Concrete Journal*, April.

CHAPTER 10

VIBRATION ISOLATED FOUNDATIONS

10.1 INTRODUCTION

For the purpose of this chapter, the vibration isolated TG foundations refer to TG tabletop structures that are supported by a vibration isolation system (VIS). This system is composed of springs and dampers, aiming at "isolating" operational machine vibrations at tabletop level from the support pedestals beneath them.

The first large turbine generator set on a spring-supported foundation was installed in the late 1960s. Nowadays, this kind of foundation can be found predominantly in Europe as well as in Asia. There is design and operating experience of numerous installations of turbine generator sets with capacities up to 1,600 MW. Most VIS-supported foundations consist of reinforced concrete frames or blocks that directly support the high-, intermediate-, and low-pressure turbines, generator, and related machine components on the tabletop. The substructure beneath the elastic support system may consist of reinforced concrete or structural steel.

While meeting the requirements of the TG manufacturer's design guidelines, a VIS-supported TG tabletop achieves operation performance levels similar to a "rigidly" supported tabletop, and in some aspects may improve its behavior and/or characteristics. Compared to rigid support structures with the same tabletop dimensions, foundation systems consisting of a VIS-supported TG tabletop may possess technical advantages based on the following aspects:

- The operational behavior of the TG set is independent of the substructure and dynamic subsoil conditions (dynamic decoupling). The substructure/subsoil needs only to resist the corresponding static loading.

- The elastic support system for turbine foundations provides vibration isolation with efficiencies higher than 98%. Adjacent structures and equipment are not affected by vibration or structure-borne noise.
- VIS provides "soft effects" resulting from differential settlement of the substructure and creep and shrinkage effects. Spring elements react smoothly when support conditions change.
- Seismic control effects can be achieved by frequency reduction of the foundation system and energy dissipation effects of dampers. Machine acceleration, relative movement, and stress or strain distribution in the structures can be reduced by a suitable choice of spring and damper characteristics.
- In the seismic case, the tabletop is not part of the lateral force resisting system, as it is connected to the substructure by springs and dampers only.
- The substructure may be designed according to regular building codes, that is, also taking into consideration cracking and plastic zones, which are generally not allowed in machine foundations.
- VIS is adjustable if system conditions are modified or updated, for instance, when seismic loading is increased as a result of modified code requirements.
- As a result of reduced seismic or dynamic loads, the dimensions of the elevated space frame pedestal foundation can be optimized (e.g., using slender columns). In this case, additional space would be available for pipework systems and/or placement of required equipment.
- The design of a TG tabletop may be standardized as the substructure and the subsoil do not influence the operational behavior of the tabletop. Modifications in the seismic input may simply result in the modification of the spring and damper design or layout.
- VIS enables straightforward renovation or retrofit of existing machine foundations.

To achieve some or all of the described advantages, however, the design engineer must have a full understanding of the behavior of all structural elements and their interactions under every possible loading condition. Besides properties of conventional steel or concrete members, the design engineer must have background information about stiffness, load, displacement, and velocity conditions of spring elements and dampers. Simplified static analysis procedures for such a TG tabletop would not predict real performance; hence, more sophisticated dynamic analysis procedures should be used for design of such foundations. The foundation design may become more complex compared with a traditional design without VIS.

When VIS is used, schedule and overall cost benefits for the execution and operation of the power plant are possible, especially when the TG substructure is integrated with adjacent building structures. The number and sizes of pedestals supporting the tabletop could also be reduced. However, if the TG pedestal foundation is integrated with the turbine building, close coordination between the TG foundation designer and the turbine building designer becomes necessary. This may add complexity to a project's execution, since the TG pedestal foundation and the turbine building may be designed by different contractors.

The following sections provide specific properties of suitable elements of VIS as well as their influence on the characteristics and behavior of the TG tabletop and its substructure under operational, emergency, and seismic loading conditions.

10.2 DESIGN PROPERTIES OF SPRING ELEMENTS AND VISCOUS DAMPERS

The performance of the TG set mainly depends on the properties of the foundation tabletop in combination with its supporting spring elements and columns, if present. Spring elements are flexible in both vertical and horizontal directions. The vertical stiffness of the springs is much lower in comparison to the vertical stiffness of supporting columns. The horizontal stiffness of the spring elements depends on the intended foundation behavior under seismic conditions.

For the design of spring-supported TG tabletops, the stiffness of the spring elements [Fig. 10-1(a)] can be described with two parameters, the spring constants k_v and k_h for the vertical and horizontal direction, respectively. Bearing capacities of a single spring element may reach nominal loads up to 1,000 kips (4,500 kN). Typically, the vertical load is achieved within the linear range, which is important when all the significant load cases are applied in the design ($F_{sv} = k_v * s_v$, where s_v is the vertical relative

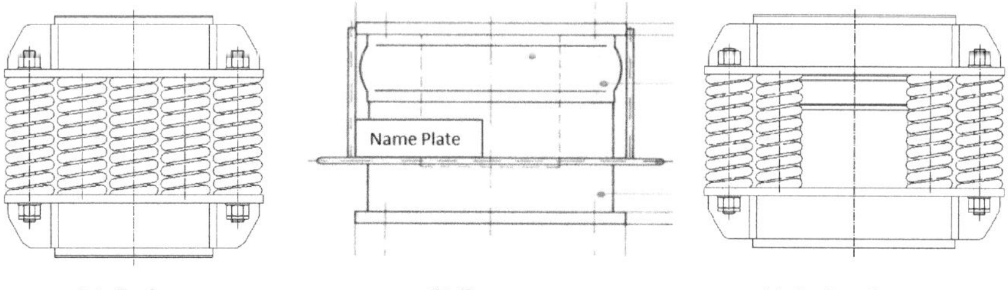

(a) Springs (b) Damper (c) Spring-Damper Combination

Fig. 10-1. Typical spring element, damper, and spring-damper combinations for TG foundations
Source: Courtesy of GERB Vibration Control Systems; reproduced with permission.

Table 10-1. Basic Design Properties for VIS

Spring elements:	$F_{sx} = k_x * s_x$	$F_{sy} = k_y * s_y$	$F_{sz} = k_z * s_z$
Damper elements:	$F_{dx} = c_x * v_x$	$F_{dy} = c_y * v_y$	$F_{dz} = c_z * v_z$

Note: x, y, z = directions; F_s/F_d = spring / damping force in kip (kN); k = stiffness value in kip/in. (kN/mm); s = relative displacement in in. (mm); c = damping value in kip*s/in. (kNs/m); and v = relative velocity in in./s (m/s).
Source: Courtesy of GERB Vibration Control Systems; reproduced with permission.

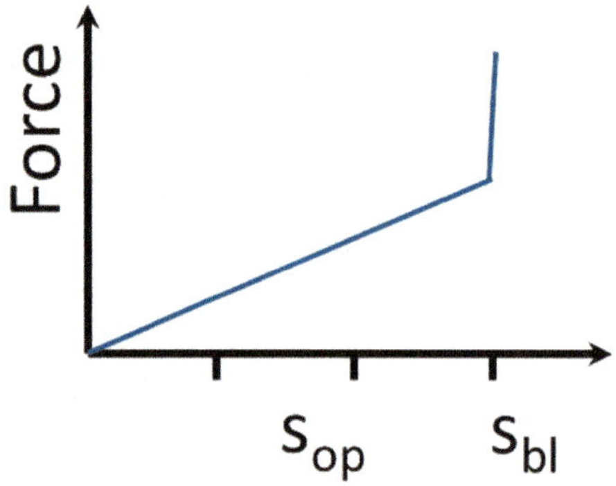

Fig. 10-2. Vertical load-displacement curve of a spring element
Source: Courtesy of GERB Vibration Control Systems; reproduced with permission.

displacement). For details, see also Table 10-1. For ultimate limit state analysis, the spring element typically possesses margins of at least 50% until the rigid height of the spring element (block height) is reached. In the block position, the vertical downward load may increase without further relative displacement of the element. The linear working range of the spring is, in most cases, sufficient for the analysis of spring-supported structures. Modeling beyond design loads may be important, for instance, in nuclear applications. For these cases the load-displacement curve may be modeled according to Fig. 10-2, where the vertical downward load increases with high stiffness beyond the block position. The working range for VIS elements used in TG foundations is typically s_{op} = ¾ in. – 1¼ in. (19–32 mm) and the block displacement s_{bl} = 1⅞ in. – 2⅛ in. (48–55 mm). More specific requirements or other load conditions should be discussed with the manufacturers of such systems.

In combination with the spring elements, viscous dampers may be used as well. Viscous dampers are dashpot elements filled with fluid of high viscosity, and are connected to the supporting component. A piston is immersed in the fluid from the top and damper forces are introduced by faster relative movement between the top and bottom of the damper [Fig. 10-1(b)]. It is obvious that the damper cannot take any dead load. Sometimes the viscous damper is integrated into the spring element ("spring-damper combination"). In this case, the fixing of the damper becomes much easier, since piston and housing are connected to the top and bottom shell of the spring element [Fig. 10-1(c)].

For design of the TG tabletop, the viscous dampers are typically represented by two constant viscous damping coefficients c_v and c_h for the vertical and horizontal direction, respectively. Damping forces are introduced by relative velocities between the upper and lower parts of the device, for example, $F_{dh} = c_h * v_h$, where v_h is the horizontal relative velocity (for details see also Table 10-1). The relative displacements of these dampers are restricted by the distance of the piston and the housing in the horizontal and vertical directions. Typical values for clearance are about 30–50 mm for operating condition. Special requirements for loading and relative displacements should be discussed with manufacturers of such systems.

For numerical modeling of spring elements, it is essential to use representative elements that do not introduce artificial effects into the model. Spring elements transfer vertical and lateral loads only. When the effect of dampers must be included, it is important to use an FE program that can appropriately model velocity-proportional dampers, or even more complex viscoelastic dampers.

F_h and F_v can be used to design the bolted connections at the VIS-substructure interface, as required (Fig. 10-3; note that $\alpha * F_h$ represents the force couple converted from the horizontal damper force, F_h). Spring elements

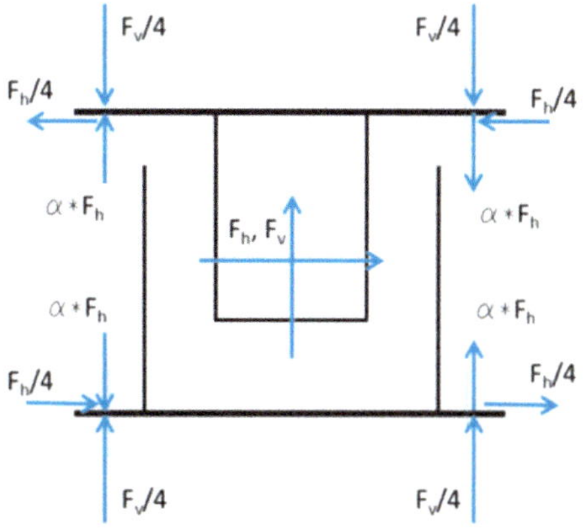

Fig. 10-3. Horizontal and vertical damper forces for design of connection bolts Source: Courtesy of GERB Vibration Control Systems; reproduced with permission.

and viscous dampers have a very small effect on rotation (rotational or torsional stiffness or damping), and can be ignored.

10.3 TYPICAL LAYOUT OF VIS FOR LARGE STG FOUNDATIONS

In a down exhaust TG arrangement, the condenser sits beneath the LP turbine, leading to a pedestal-tabletop type of foundation. Fig. 10-4(b) shows a possible layout for a VIS-supported STG foundation in comparison with a traditional STG pedestal foundation [Fig. 10-4(a)]. A traditional STG pedestal foundation is physically isolated from adjacent foundations to avoid vibration transmissions, whereas for a VIS-supported STG foundation the basemat can be integrated with other adjacent foundation mats, which may offer significant advantages in areas with groundwater influence.

The spatial distribution of the spring elements is chosen in such a way that the vertical spring displacement is almost constant in every element under operating loading conditions. The layout of the substructure plays an important role here, especially the supporting columns and beams. A typical structural layout of the substructure is given in Fig. 10-5.

10.4 DYNAMIC CHARACTERISTICS OF ELASTICALLY SUPPORTED TG FOUNDATIONS

Foundation systems consisting of VIS-supported TG tabletops exhibit different characteristics in natural frequencies, mode shapes, and damping ratios from traditional TG foundations. Independent of the supporting

(a) Traditional Pedestal Foundation Layout (b) Spring-supported Pedestal Foundation Layout

Fig. 10-4. Typical cross section of machine buildings, without and with VIS; (a) traditional pedestal foundation layout; (b) spring-supported foundation layout
Source: Courtesy of GERB Vibration Control Systems; reproduced with permission.

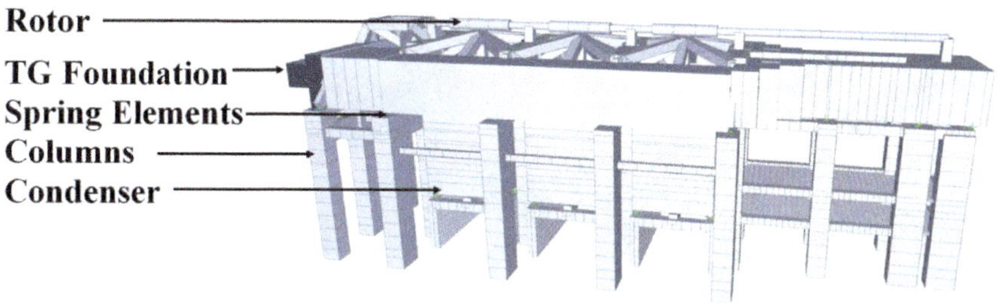

Fig. 10-5. Typical analytical model for spring-supported STG tabletop with attached condenser
Source: Courtesy of GERB Vibration Control Systems; reproduced with permission.

Fig. 10-6. Typical rigid body mode in the longitudinal direction (tabletop)
Source: Courtesy of GERB Vibration Control Systems; reproduced with permission.

columns, there are six natural frequencies in the range between 1 Hz (horizontal) and 5 Hz (vertical), which correspond to the six rigid body modes of the TG tabletop itself. Depending on the flexibility of the supporting columns, the machine-foundation system frequencies in horizontal directions will be further decreased, since the column lateral stiffnesses may be in the same order of magnitude as those of the springs (Fig. 10-6). However, there is insignificant change in the vertical frequencies, because the vertical stiffnesses of the supporting columns are much higher than those of the springs. The combined stiffness of the machine foundation system can be calculated from the stiffness of the spring system in series with the stiffness of the supporting column system, which may consist of reinforced concrete or structural steel.

Dampers provide local damping effects, and are usually placed next to spring elements. The overall damping ratio for the foundation system may be significant, as the spring systems represent the most flexible part of the support system. Typical damping ratios of the rigid body modes

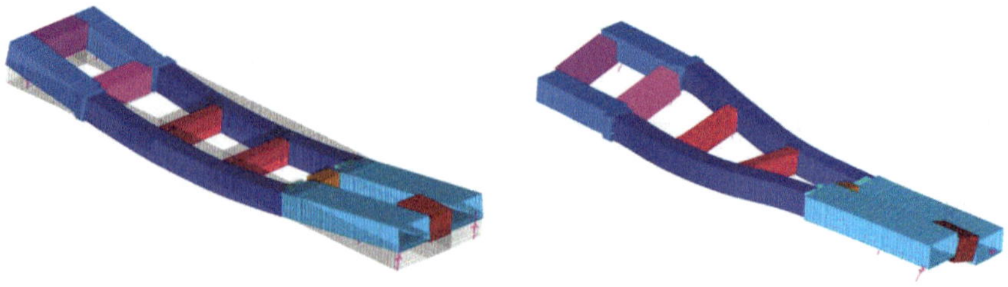

Fig. 10-7. Primary bending and torsional modes of spring-supported STG table-tops (frequency range II). See Table 10-2
Source: Courtesy of GERB Vibration Control Systems; reproduced with permission.

(first six natural frequencies) are usually in a range between 5% and 30% of the critical damping value. For low/moderate seismic zones, the damping ratios can be designed to lower values, while in high seismic zones, the desired damping ratio may reach 30%, especially in horizontal directions.

Primary bending and torsional frequencies of TG tabletops are usually in the range between 4 Hz and 8 Hz, depending on the mass (distribution) of the tabletop and machinery, as well as on dimensions and the stiffness (distribution) of the tabletop itself (Fig. 10-7). The spring elements and their positioning, as well as the properties of the substructure, have only minor influence on such frequencies.

Tabletop modal frequencies close to operating speed (Fig. 10-8) usually have very low energy contents (i.e., these modes have very low mass participation factors, for example, less than 0.1% or even less), thus the foundation "resonance effects" at operating speed will be minimal. Damping ratios for these modes usually correspond to the damping of the foundation material, which is 2% for reinforced concrete foundations.

Table 10-2 shows typical frequency ranges for a spring-supported STG tabletop with operating frequency of 30 Hz.

10.5 OPERATIONAL BEHAVIOR

Because the fundamental frequencies for spring-supported STG tabletops (i.e., rigid body modes) are far from machine operational excitation frequency at 25 Hz, 30 Hz, 50 Hz, or even 60Hz, there should not be any danger of vibration amplifications in the operating speed range for an appropriately designed STG tabletop supported by a VIS. There are only limited effects from the natural frequencies (and corresponding mode shapes) close to the excitation frequency because of their low energy contents (Fig. 10-8).

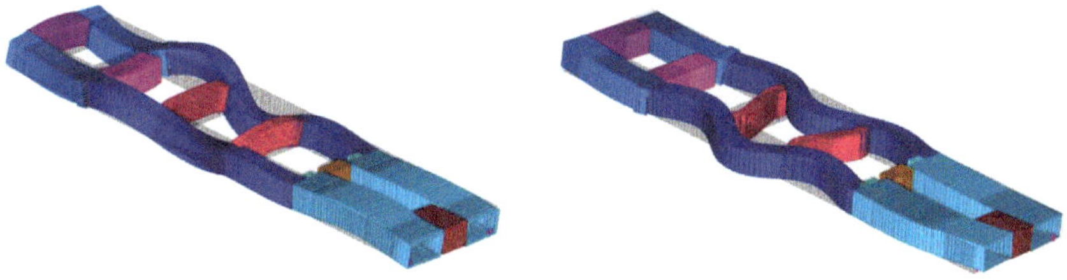

Fig. 10-8. Higher modes close to operating speed (frequency range III)
Source: Courtesy of GERB Vibration Control Systems; reproduced with permission.

Table 10-2. Typical Frequency Ranges of Spring-supported STG Tabletops with Operating Speed of 30 Hz

Range	Mode Order	Frequency	Description
I	1–6	1–5 Hz	Rigid body modes
II	7–x	4–27 Hz	Bending and torsional modes below operating speed range
III	$(x+1)-y$	27–33 Hz	Higher bending and torsional modes in the operating speed range
IV	$y+1$	>33 Hz	Modes beyond operating speed range

Source: Courtesy of GERB Vibration Control Systems; reproduced with permission.

Some typical results of a forced vibration analysis are shown in Fig. 10-9. Certain natural frequencies near the operating speed (30 Hz) may contribute to the peaks shown in the frequency range of interest. The vertical and horizontal stiffness of the VIS does not have a significant influence on the amplitudes at operating speed. The frequencies of rigid body motion (around 1 Hz–5 Hz) are clearly out of the avoidance range, which is 30 Hz +/− 5 Hz in the given example. The effects of dampers at operating speed can also be neglected and the structural damping is assumed to be 2% for concrete foundations. Typical amplitude limits of machine manufacturers or international standards can be achieved. For example, per ISO 20816-2, the bearing vibration amplitude limit (0 to peak) for new

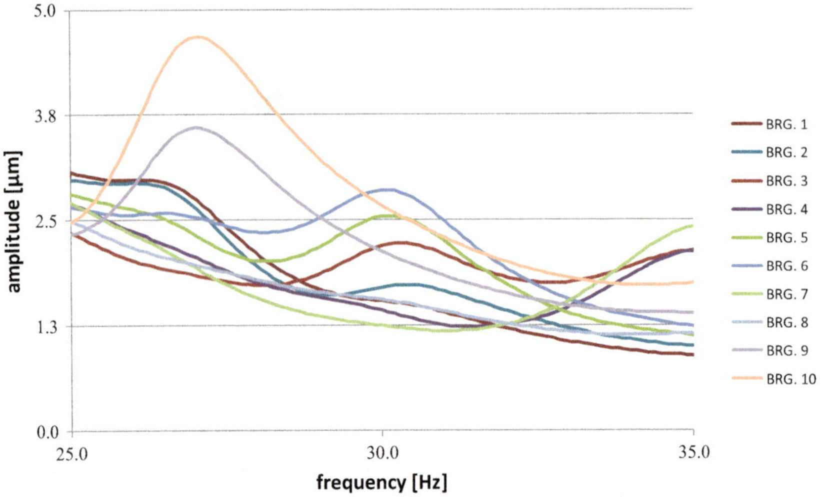

Fig. 10-9. Expected 0-peak vibration amplitudes at 10 bearings at operating speed 30 Hz
Source: Courtesy of GERB Vibration Control Systems; reproduced with permission.

STG machinery is 17.1 μm at 50 Hz [(3.8 mm/sec) / (2*3.14*50) * 1.414)]. Additional details in regard to the analysis of vibration isolated TG decks is given in Hüffmann et al. (2008).

During startup or shutdown of the turbine set, there are many resonance peaks in between (Fig. 10-10) due to machine speeds temporarily matching foundation natural frequencies along the way. However, the unbalance forces at low frequencies are small and thus, even in the area of the low speed resonances, the expected amplitudes are acceptable. Damping effects of the viscous dampers also help reduce the amplitude levels, especially for the rigid body frequencies (during startup or shutdown), and can be combined with material damping ("composite modal damping") in the analyses.

For the proper operation of the shaft system, a certain dynamic stiffness value in the bearings has to be provided in the range of the operating speed. Typical minimum values are between 2 MN/mm and 4 MN/mm. Some of the machine manufacturers have specific requirements. A typical plot of expected stiffness in the frequency domain of interest (operating speed 30 +/− 5 Hz) is given in Fig. 10-11. These curves are derived from steady state response analyses in the corresponding frequency domain. The bearing stiffness of TG foundations with VIS is typically well higher than the required values.

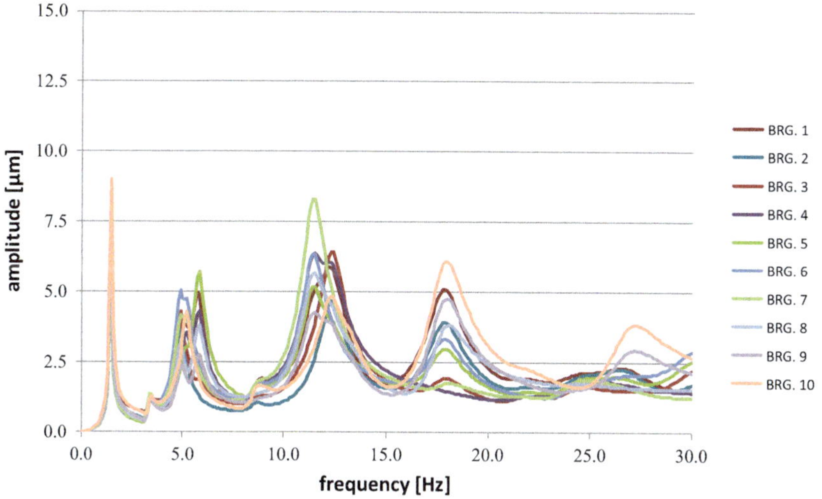

Fig. 10-10. Typical startup or shutdown amplitudes of a 1,000 MW STG machinery at operating speed 30 Hz
Source: Courtesy of GERB Vibration Control Systems; reproduced with permission.

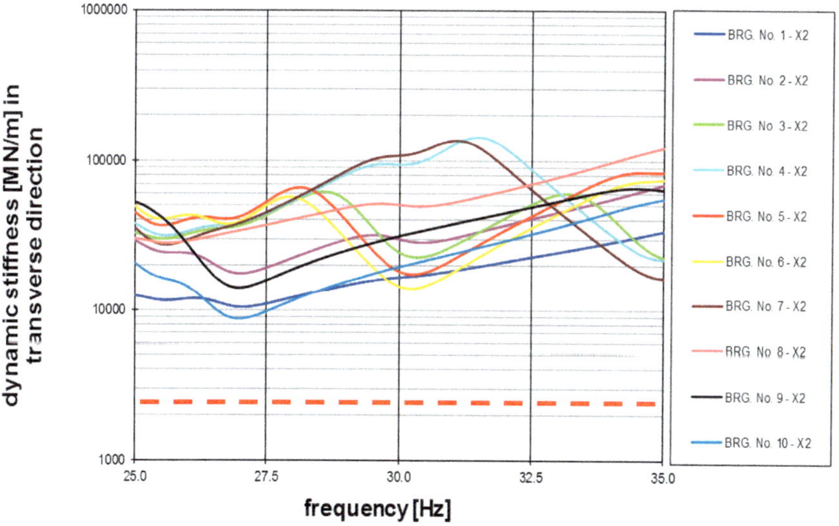

Fig. 10-11. Typical dynamic bearing stiffness around operating speed 30 Hz
Source: Courtesy of GERB Vibration Control Systems; reproduced with permission.

10.6 EMERGENCY CONDITIONS

Spring-supported STG tabletops "appear" quite flexible due to the characteristics of the springs. However, such tabletops do have significant stiffness and mass; thus the deflection for an appropriately designed tabletop would be very small. For instance, assume a STG tabletop with a total mass of 8,000 metric tons, including machinery. The vertical stiffness of VIS is usually designed to deform no more than one inch (25.4 mm) under this self-weight ($K_v = 8{,}000 * 9.81/25.4$ mm = 3,139 kN/mm), and the horizontal stiffness of the VIS is usually designed as 50% of the vertical stiffness (3,139/2 = 1570 kN/mm). This means that a 1570 kN static force would be required to move the tabletop by just 1 mm in the horizontal direction.

With the previous discussion in mind, it is recognizable that even emergency conditions of the machinery will not impose severe damage to the system. As an illustration, a time history dynamic analysis for a generator short circuit load is performed, and the relative displacements at the bearings are shown in Fig. 10-12. The peak vibration is 1.1 mm. For extreme cases, 2.0 mm can be expected. 1–2 mm can also be expected as additional relative displacement in the VIS. This level of displacement is well within the safety margins of the system. Note that there is no uplift of the springs in the vertical direction, as the tabletop and machinery self-weight is large enough to overcome the short-circuit uplift effects.

10.7 SEISMIC CONSIDERATIONS

In areas of medium to high seismicity, the overall frequency and damping behavior as well as the corresponding mode shapes play very important roles for the performance of a spring-supported tabletop. By varying the properties of the VIS, it is possible to effectively control the foundation's seismic responses. In high seismic areas, it is common to choose horizontally flexible spring elements with dampers with a higher viscous damping coefficient. Such tunings to gain better seismic control have no effect on the operational behaviors of the TG, and only limited influence on emergency conditions of the machinery.

For configurations with springs according to Fig. 10-4, typical foundation fundamental frequencies are 0.7 Hz to 1.0 Hz in the horizontal direction. Meanwhile, the damping ratio for the relevant mode(s) may reach 20% of critical damping. Therefore, it is expected that the amplification of seismically induced acceleration will be quite low, and in some cases complete "isolation effects" can be achieved. This can be illustrated in a typical response plot of peak acceleration along the height of the foundation without and with a suitable VIS. As shown in Fig. 10-13, the seismic acceleration amplification can be at the peak for a rigidly connected STG tabletop structure,

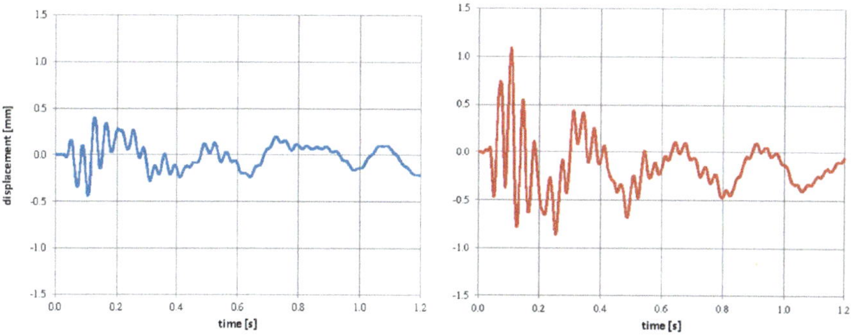

Fig. 10-12. Horizontal (left) and vertical (right) displacements at a generator bearing by generator short circuit
Source: Courtesy of GERB Vibration Control Systems; reproduced with permission.

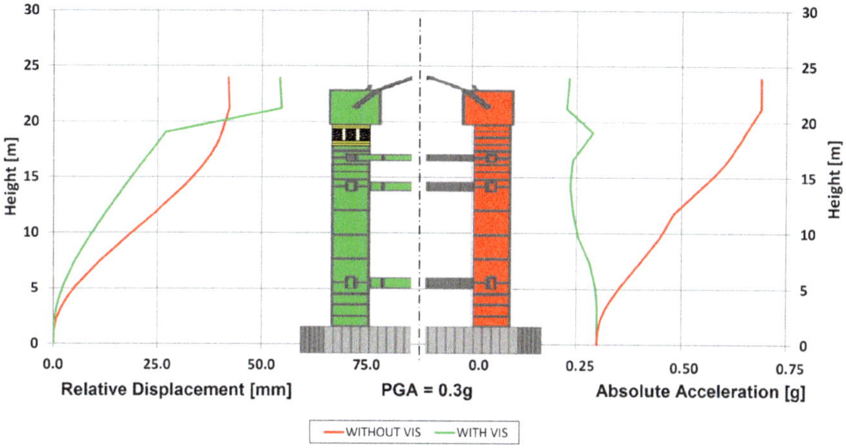

Fig. 10-13. Seismic response of TG foundation (without and with VIS); peak displacements and acceleration along height
Source: Courtesy of GERB Vibration Control Systems; reproduced with permission.

while a VIS-supported STG tabletop can lead to significantly lower levels of acceleration, especially at the turbine level. The left side of Fig. 10-13 shows the corresponding plot of peak displacements along the height of a TG foundation. Typical cases for a peak ground acceleration (PGA) of 0.3g lead to relative displacements in the VIS of about 25–50 mm (1 in.–2 in.). The actual amplitude depends on given seismic conditions as well as on the substructure and stiffness/damping properties of the VIS. More information of an executed project is given in Nawrotzki et al. (2010).

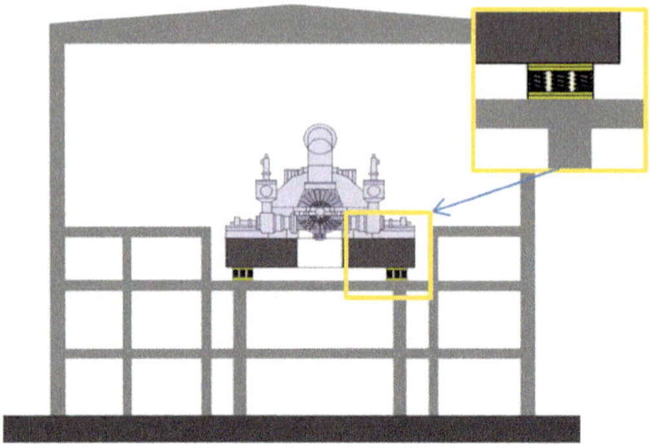

Fig. 10-14. Spring-supported STG tabletop integrated in a machine building
Source: Courtesy of GERB Vibration Control Systems; reproduced with permission.

The substructure of the spring-supported STG tabletop may also be integrated into the machine building consisting of steel or concrete (Fig. 10-14). The layout of the machine building is not restrained by any interaction effects, as vibration and noise are suppressed by the VIS. Column rows may be eliminated as the columns supporting the TG tabletop may also carry the loads from adjacent floors. The space beneath the TG tabletop becomes more open, which has significant advantages for the arrangement of equipment, as well as for the installation of piping.

In configurations similar to Fig. 10-14, the design of the machine building structure has to take into account the parameters of the STG tabletop, including VIS. The spring-supported mass is significantly high, and hence the STG tabletop possesses a measurable influence on the seismic behavior of the machine building. Since the tabletop is connected flexibly and damped to the rest of the structure, the effects of the tabletop ("mass damper effects") reduce the seismically induced motion of the machine building. Further characteristics of an integrated structure are

- reduction of induced machine acceleration;
- reduction of relative movement between substructure and machine building (Fig. 10-14); and
- reduction of structural stresses in substructure and machine building.

These effects are described in more detail in Basu et al. (2011) for a 660 MW project.

Spring elements and dampers can be classified as "structural elements," since they are vertical and horizontal load bearing components, and in most cases are integrated with the structural steel or reinforced concrete surroundings. The foundation designer should clearly identify load transfers between the structure and springs/dampers at the top and bottom of the VIS. The proof of stiffness and strength of the connections according to actual building codes is the responsibility of the foundation design engineer. Manufacturers of the VIS may provide assistance in the corresponding design approaches.

It is recommended that the response modification coefficient *R*, as given in ASCE 7, Chapter 15, for a VIS-supported concrete tabletop structure, be taken as 1.0 or 1.25, to minimize plastic zones and cracks in concrete that may lead to malfunction of the machinery or the VIS connections to the substructure. Consequently, the same *R*-value should also be used for the design of the spring elements, as these elements should behave elastically up to the seismic design displacements. Damping may reduce the demands in terms of relative displacements in the devices as well as stresses and strains in the adjacent structures, and should be included in the seismic analysis and design.

ASCE 7, Sections 12.2.3.1 and 12.2.3.2, can be used to determine the seismic design parameters for a TG foundation with VIS, as applicable. Use of an *R*-value higher than 1.25 can be allowed for the strength design of the substructure beneath the VIS.

In general, however, ASCE 7, Chapter 17, is not applicable for VIS-supported TG tabletop structures in the scope of this chapter, since an elastically supported STG tabletop is not a "horizontally flexible and vertically stiff structural element." The structural design of the tabletop should follow ASCE 7, Chapter 15, and other general code provisions, as well as the STG manufacturer's special requirements.

The elastic support devices and the load transfer mechanism (connection between VIS and structures) have to be designed to withstand load and displacement combinations according to all applicable requirements.

Prequalification of elastic support devices as well as quality control requirements must be discussed for each specific project. The fulfillment of environmental conditions, creep, shrinkage, and temperature effects must be addressed for every application, as applicable. The possibility for inspection, maintenance, and replacement must also be discussed with the general contractor and owner(s).

10.8 SPRING-MOUNTED CONDENSERS

For steam turbines, different condenser arrangement in the LP turbine area may impose different loading pattern onto the foundation. In most cases (>100 MW), the condenser sits beneath the LP turbine leading to a tabletop type foundation (see Chapter 3). One possible design could be that the condenser is welded to the LP turbine nozzle with a smaller part of the condenser weight, typically 20%–30%, hanging at the LP turbine (Fig. 10-15, left side), while the remainder of the condenser dead load is supported by

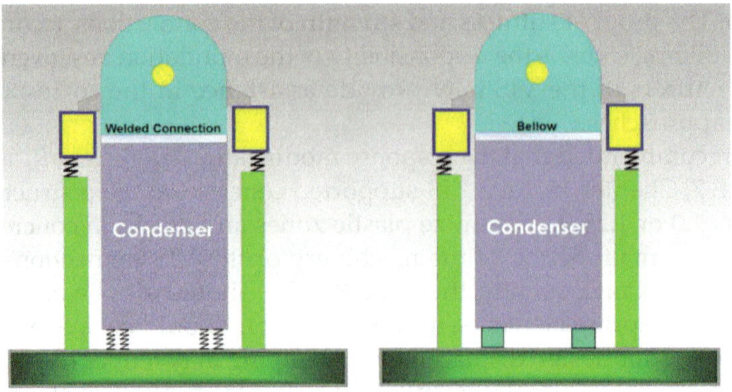

Fig. 10-15. Typical configurations of spring-supported and rigid condenser Source: Courtesy of GERB Vibration Control Systems; reproduced with permission.

springs, which allows heat expansion of the condenser downward, as the condenser springs are much softer than the spring system supporting the STG tabletop. The advantage of this arrangement is the elimination of the vacuum pull (vacuum load, see Chapter 4), which represents a potential danger in regard to the leveling of the shaft system. The vacuum pull becomes an internal force in the condenser shell in the case of a spring-supported configuration. In this case, only the hanging part of the condenser has to be considered in the foundation design.

In an alternative arrangement, the condenser is fixed at the bottom with a bellow between the condenser and the LP turbine nozzle to allow heat expansion upwards (Fig. 10-15, right side). In this case, neither dead load nor heat expansion will be part of the foundation analysis. However, the vacuum pull becomes an external load to the foundation. This configuration may require a stiffer spring system in the vertical direction to satisfy the static stiffness criteria also under full vacuum pull. Typical spring stiffness values result in vertical displacements of the TG tabletop springs under operational load of about 10 mm to 30 mm ($f = 5$ Hz/2.8 Hz). For a quick estimation of the vertical rigid body frequency, a simplified formula, $f = 5/\sqrt{s}$, with the static displacement s in cm, can be used.

10.9 GAS TURBINE FOUNDATIONS

Gas turbine foundations may also be equipped with VIS, especially when there is a danger of differential soil settlement or high seismic risk. Typical support frequencies can be chosen between 2.5 Hz and 5 Hz. There is no deviation of the specification of the machine manufacturer, as all the requirements can be satisfied also as a spring-mounted foundation. The main advantages of VIS given in Section 10.1 are also valid for combustion turbine foundations.

10.10 RETROFIT OF TURBINE FOUNDATIONS

Many power plants have been in operation for more than 30 or 40 years. In some of these cases, an update of the machinery becomes necessary to increase capacity or efficiency or to meet required emission standards. Spring elements have been successfully used for the retrofit of existing turbine foundations.

One retrofit approach is to remove the whole TG tabletop and parts of the columns. On top of the prepared column heads, spring elements are installed, which also provide part of the support for the construction of a new tabletop. Another approach is to keep the original tabletop and cut out suitable parts of the columns (Fig. 10-16). Prestressed spring elements are installed in the corresponding vertical clearance, which are then released and adjusted (also see Section 10.12).

10.11 SELECTIVE APPLICATION EXAMPLES

There are more than a thousand operating spring-supported TG tabletops worldwide. The range of capacities varies from 10 MW to 1,600 MW and the steam may be generated conventionally or by nuclear technologies. One of the first installations was in operation beginning in the 1970s (Stade Nuclear Power Plant, Germany). VIS technology has successfully been used on turbine tabletops for over 40 years.

Fig. 10-16. Typical steps for retrofit of a TG tabletop
Source: Courtesy of GERB Vibration Control Systems; reproduced with permission.

Fig. 10-17. 660 MW turbine tabletop on VIS integrated into machine building, Jhabua, India
Source: Courtesy of GERB Vibration Control Systems; reproduced with permission.

Some details of a 660 MW TG foundation (Jhabua Power Plant, India) are given in Fig. 10-17. The spring-supported TG tabletop is sitting on steel columns that are integrated in the machine building. The spring elements are especially flexible in the horizontal direction and respond at about 2.8 Hz in the vertical direction. The horizontal damping ratio is quite high to accommodate the high seismic risk. Due to the reduced structural frequencies and higher damping, it was possible to reduce the steel cross sections for the machine building structure. The integration also allowed for relocating the horizontal load transfer braces to noncritical or nondisturbing locations in the plant.

Fig. 10-18 shows spring elements and spring-damper combinations, which are covered to protect them against dirt and dust during the construction phase (Ling Ao, Phase II, 2 × 1000 MW, China). The concrete columns were designed to also support some beams that are parts of the machine building (corbels in Fig. 10-18).

10.12 SPECIAL CONSIDERATIONS FOR TG FOUNDATIONS WITH VIS

Strength Analysis—The foundation designer should bear in mind that the springs/dampers are reacting according to real loads, not to (theoretical) design loading. In some cases (e.g., in high seismic zones), using the

Fig. 10-18. Ling Ao Phase II, 1000 MW turbine tabletop on VIS, China
Source: Courtesy of GERB Vibration Control Systems; reproduced with permission.

equivalent static seismic design approach, as well as equivalent static loads for machine emergency loads (e.g., generator short circuit), may result in unrealistically high spring forces and displacements. In these cases, it may be necessary to use more realistic analysis approaches, for example, based on dynamic analysis of emergency conditions (in time domain), for the design of the substructure, while the tabletop is still designed using the equivalent static approach.

Static Stiffness Requirements—The static stiffness requirements have to be evaluated under a careful choice of load cases and combinations. The springs may theoretically react excessively when unjustified operating loads are applied; for instance, a simplified static unbalance load may only be used for the reinforcement design of the STG tabletop. It should not be used to check the spring compression; otherwise it would result in unrealistic results.

Installation—There are many possible ways to install the spring elements. The installation method should be discussed between VIS supplier, foundation designer, and construction company, and be adjusted to local requirements. A standard process described here illustrates some of the features of a typical VIS installation (Fig. 10-19).

1. Prior to the first setting, the spring elements are prestressed to the height or load of the intended final situation (hydraulic, at the manufacturer or on site). The spring elements are placed in position on top

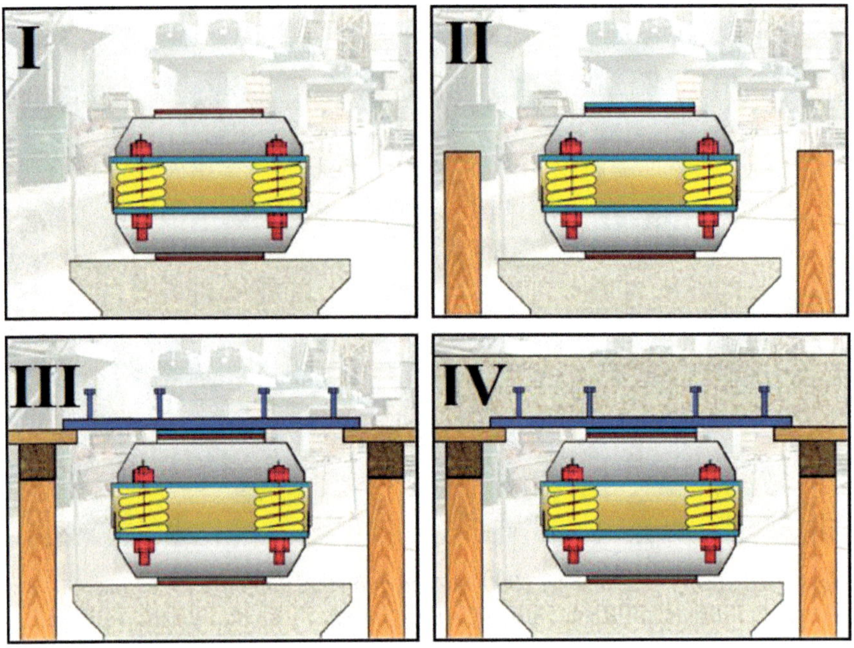

Fig. 10-19. Installation process of a standard spring system beneath a concrete tabletop
Source: Courtesy of GERB Vibration Control Systems; reproduced with permission.

of columns usually using bearing pads as a measure against an uneven surface (Fig. 10-19, I).

2. Height adjustment shims are placed on top of the prestressed devices to achieve the intended level of the TG tabletop bottom (Fig. 10-19, II).
3. Struts and shuttering are placed around the VIS elements and the openings to the spring elements, covered by bearing pads again as well as an anchored steel plate to close the opening in the formwork (Fig. 10-19, III).
4. Placement of reinforcement and concrete may go on as usual (Fig. 10-19, IV).

Maintenance—Typical spring and damper elements are maintenance-free. As the housings consist of steel, it is recommended to check the corrosion protection at regular time intervals, for example, every fifth year. If dampers are used, the filling condition of the viscous liquid should be checked; the recommended time interval is 5 years.

Leveling—After finishing the concrete installation, hardening of the concrete, and installation of all machine parts on top of the foundation, it

Fig. 10-20. Height adjustment is possible, if required
Source: Courtesy of GERB Vibration Control Systems; reproduced with permission.

may be leveled again before the springs are released (Fig. 10-20). The prestressing bolts may be removed or stay in place if they do not disturb the operations or seismic movement. Hydraulic equipment can be used to prestress the spring elements further to provide a gap at the top of the element in order to increase or reduce the height of the adjustment shims before the free springs take over the dead, operational, and all other loads acting on the devices according to design.

REFERENCES

ASCE. (2010). "Minimum design loads for buildings and other structures." *ASCE/SEI 7-10*, Reston, VA.

Basu, A. K., Nawrotzki, P., and Siepe, D. (2011). "Spring supported turbogenerator with steel supporting columns integrated into the steel structure of the turbine building." *Conference, Power-GEN India and Central Asia*, New Delhi.

Hüffmann, G., Nawrotzki, P., and Uzunoglu, T. (2008). "Static and dynamic analysis of concrete turbine foundations." *Struct. Eng. International*, 3, 265–270.

ISO (International Organization for Standardization). (2017). "Mechanical vibration—Measurement and evaluation of machine vibration—Part 2: Land-based gas turbines, steam turbines and generators in excess of 40 MW, with fluid-film bearings and rated speeds of 1,500 r/min, 1,800 r/min, 3,000 r/min and 3,600 r/min." *ISO 20816-2*, Geneva, Switzerland.

Nawrotzki, P., Siepe, D., and Basu, A. K. (2010). "Design of turbine foundations in seismic zones." *Conference, Power-GEN India and Central Asia*, Mumbai.

CHAPTER 11
CONSTRUCTION CONSIDERATIONS

11.1 INTRODUCTION

Constructing a reinforced concrete TG foundation is one of the major tasks at a power plant construction site. This is so not only because of the required amount of concrete and reinforcement compared with other concrete structures, but also because of the complexities associated with the different types of embedded items and associated interferences with reinforcements, equipment installations, and long construction lead times in the schedule.

ACI 351.3R, Chapter 5, addresses construction considerations for dynamic equipment foundations, including TG foundations in general. The contents of ACI 351.3R, Chapter 5, are not repeated here. Only specific construction issues unique to TG foundations are presented in this chapter.

11.2 CONSTRUCTION JOINTS

It is desirable to construct any structure using the largest concrete placements possible with the fewest construction joints. However, consideration must be given to the feasibility of such large placements. For TG foundations, construction joints are required between the top of the basemat and the columns, walls, piers, and/or blocks, between the top of the columns and the tabletop, and sometimes within the basemat as well.

In the basemat, horizontal construction joints at the middle surface of the basemat, or vertical construction joints at locations with lower bending moments, have been used. For vertical construction joints in the basemat, horizontal dowels can be added across the joints. The design engineer can use shear friction provisions in ACI 318 to conservatively determine the required dowels to provide a capacity equal to the concrete out-of-plane shear capacity. Engineering notes should be added to the drawings to provide information on the appropriate surface preparation at the joints per applicable code requirements. Horizontal construction joints are not recommended, as they usually require extensive cold joint preparation work on too large an area to be economical.

For axial-exhaust TG foundations, construction joints should not be allowed within piers and columns where the vacuum load is resisted. The large vacuum load may introduce undesired deflections because of reduced stiffness at the construction joints if not prepared and constructed appropriately.

11.3 CONSTRUCTION TOLERANCES

In addition to the construction tolerance requirements per applicable codes (e.g., ACI 318), vendor-specific tolerance requirements on anchor bolts and special embedded items should be included on the construction drawings. These requirements are critical for successful installation of the heavy TG equipment.

Special structural templates are required to secure the embedded items during the concrete pour. An example of a structural template is shown in Fig. 11-1.

11.4 CONCRETE PLACEMENT

11.4.1 Congestions

Reinforcing bar congestion is a major issue in TG foundations because of the large size of the structural members, heavily reinforced sections, and the large number of embedded items required by equipment manufacturers. In many instances, the geometry of the members controls the amount of reinforcing required. This includes minimum flexural reinforcement as well as minimum torsional reinforcement; both hoops and longitudinal reinforcing add to congestion. Reinforcement interrupted at embedded mechanical or electrical commodities must be accounted for in the design and detailing stages.

Headed reinforcement can be utilized to limit congestion at joints in TG foundations. The use of headed reinforcing bars eliminates the need for hooked bars and reduces the embedment length (Fig. 11-2). Additional

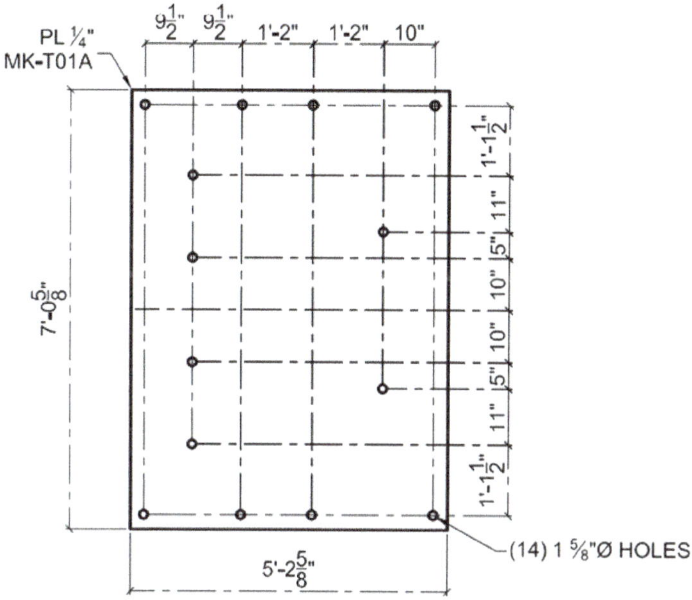

Fig. 11-1. Example of large bolt group template

Fig. 11-2. Example where headed reinforcement is used in congested joint
Source: Courtesy of Kiewit Engineering Group; reproduced with permission.

design checks are required when using headed reinforcing bars at member edges and should be evaluated by the design engineer. Mechanical rebar couplers can also be utilized at congested areas where rebar splices are excessive.

Because of reinforcement congestion, it is important that the design engineer specify the concrete mix to be used for a TG foundation on the construction drawings. Using the right concrete mix will eliminate potential concrete placement issues such as honeycombs. For TG foundations, the congested areas are at the column-tabletop interfaces, beam intersections, large embedments, pipe sleeve penetration locations, and machine anchor bolt locations. Three-dimensional computer detailing programs are capable of modeling and detailing congested areas, resolving and reducing potential issues prior to construction.

For TG foundations, the design engineer should specify a concrete mix with a large slump and smaller coarse aggregates such as pea gravel for congested areas. To be economical, different concrete mix types may be specified for different concrete elements in a TG foundation. For example, the design engineer may specify a lower slump with larger aggregate mix for large volume pours such as the basemat, columns, pedestals and walls, while specifying a higher slump with smaller aggregate mix for more congested elements such as the TG tabletop. When considering the use of specialty mixes for massive concrete placements, it is important to consider the effects of the high heat of hydration. These can be addressed by considering the type of aggregate and class of fly ash being used.

In hard-to-reach areas and areas where reinforcing congestion cannot be reduced, self-consolidating concrete (SCC) can be used. If SCC is used for any portion of the TG foundation, additional design provisions need to be taken into account when designing the soffit and edge form support.

11.4.2 Mass Concrete Temperature Monitoring

Per ACI 301, placement of structural concrete with a minimum dimension equal to or greater than 4 ft should be considered mass concrete. Thus, TG foundations are typically mass concrete structures. Per ACI 207.2R, "the most important characteristic of mass concrete that differentiates its behavior from that of structural concrete is its thermal behavior." An important part of planning for a large mass concrete placement is a properly designed concrete mix that takes into account the specific mix temperature rise and composition. During the construction stage, a concrete temperature monitoring scheme is required to help assess and adjust needed protections while curing. A properly placed and protected mass concrete will have minimal thermal cracking during curing.

In placing concrete for TG foundations, it is recommended to monitor the concrete temperatures per the following ACI 301 requirements:

1. The maximum temperature in concrete after placement shall not exceed 158°F.
2. The maximum temperature difference between center and surface of placement shall not exceed 35°F.

Maximum temperature is limited to minimize future durability concerns because of delayed ettringite formation (DEF) and potential reductions in ultimate strength, whereas the maximum temperature difference is limited to minimize thermal cracking. Depending on specific mix design, placement dimensions, and reinforcement configuration, other temperature limits may be possible.

A thermal monitoring scheme should be prepared per ACI 301 requirements. A drawing should be issued to Construction to show placement of thermocouples. Per ACI 301, a temperature sensor should be placed at the center of the mass of placement, and another at a 2 in. depth from the top of the concrete surface at the center of the pour. Backup sensors should be provided in case other thermocouples fail or are damaged during concrete placement.

It may be necessary to consult with a concrete material specialist to plan for a successful mass concrete placement.

11.5 MODULARIZATION

Early involvement of the construction contractor in the TG foundation design is necessary if modular reinforcement cages are to be used. The construction of large reinforcement cages in a fabrication shop or at grade allows the construction contractor to install the cages and reduce the amount of work at heights. Guidelines for rebar cage safety can be found in Casey and Urgessa (2013).

The column rebar cages are an example of a component that can be modularized (Fig. 11-3). Rebar cages for tabletop beams, both longitudinal and transverse, can also be pre-tied and placed as modules, which requires additional coordination with the construction contractor to detail the reinforcement. Individual beams can be modularized, or all four beams around a tabletop opening can be tied together and lifted into place.

For relatively small to medium size TG foundations, precast concrete can be an attractive alternative for columns and tabletop, when shortening the construction schedule becomes critical. The columns can be fabricated in segments and then assembled at the site. The bottom portion of tabletop can be prefabricated and serve as permanent formwork for top

Fig. 11-3. Column rebar cage being flown into place
Source: Courtesy of Kiewit Engineering Group; reproduced with permission.

reinforcement. Concrete is then cast in place to complete the tabletop structure.

11.6 POST-INSTALLED ANCHORS AND INTERFERENCE WITH REINFORCEMENT

After the STG tabletop concrete is placed and cured, post-installed anchor bolts (mechanical or adhesive) are used to secure turbine or generator enclosures, handrails, and other miscellaneous support items on the STG tabletop. The construction contractor typically encounters reinforcing bars while installing these anchor bolts. After consultation with the design engineer, the contractor may not be allowed to cut certain reinforcing bars. For locations where rebar cutting is not allowed, the construction contractor will need to map existing rebar to determine the locations to drill new holes and fill in abandoned holes, and may even have to alter the enclosure structure to accommodate the revised bolt plan, causing additional cost and schedule implications.

In practice, the machine supplier usually subcontracts the enclosure design, fabrication, and delivery to a lower-tier, specialty contractor. This contractor may not be engaged early enough for the anchor bolt locations to be available for incorporation into the tabletop design drawings, causing the interference issues identified previously.

There are options available to minimize or eliminate such issues. The construction contractor and design engineer should communicate closely before deciding which option to use for a particular project.

11.6.1 Cast-in-Place Option

If the vendor information on these support items is available during the design stage, design engineers can specify anchors with coupling nuts as shown in Fig. 11-4 to eliminate this issue.

11.6.2 Post-Installed Option

This solution provides the most flexibility to the construction contractor, and is considered a better solution than the cast-in-place option for miscellaneous smaller anchor bolts (e.g., machine enclosures). Expansion anchors, epoxy anchors, or cementitious grout anchors can be used. This may be the only option when the information is not available when placing the tabletop concrete.

For this option, at the design stage, the design engineer can specify all tabletop main reinforcement (flexural and torsional/shear ties) at least 6 in. below the tabletop concrete and provide temperature reinforcement (e.g., #4@12 in. or #5@12 in.) 1.5 in. to 2 in. below the concrete surface. Temperature reinforcements may be cut during the post-installed anchor installation, as these reinforcing bars are not required for strength. Placing

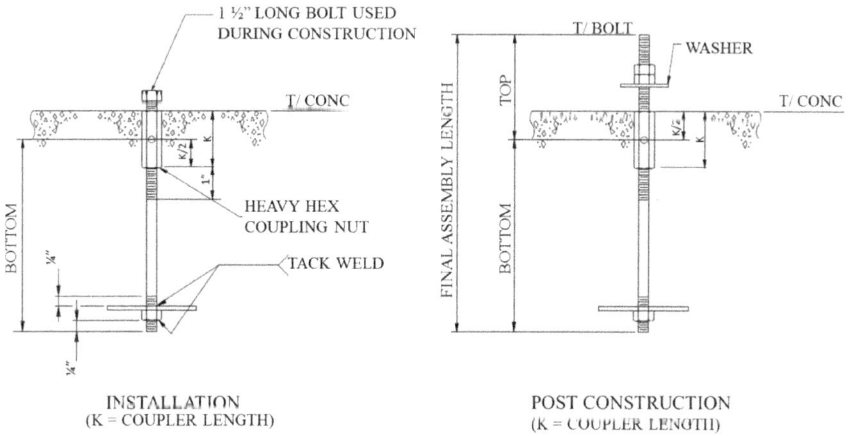

Fig. 11-4. Typical anchor with coupling nut (coupler)

the main reinforcement 6 in. or more below top of concrete should not cause any design issues, as these structural members are, in general, oversized because of stiffness considerations.

11.7 TEMPORARY CONSTRUCTION LOADS

In addition to the permanent loads induced by the machine, there may be temporary loads applied to the tabletop because of the construction contractor's preferred method of installation. A gantry system that is supported on the tabletop may be used, as shown in Fig. 11-5. Close coordination with the rigging contractor is required to identify the location and magnitude of these temporary support loads. ASCE 37 can be utilized to estimate construction loads as required.

11.8 CONDENSER INSTALLATION

Because of its large size and connection configurations with the low-pressure (LP) turbine exhaust hood and limited available space, installation

Fig. 11-5. Generator installation with gantry system supported off tabletop
Source: Courtesy of Kiewit Engineering Group; reproduced with permission.

of a bottom exhaust condenser requires close coordination between design engineer, condenser manufacturer, TG manufacturer, and construction contractor. See Figs. 11-6 and 11-7 for example illustrations.

The welds between the LP hood and the condenser flange are performed from within the condenser. If not, notches or block-outs may be required in tabletop concrete members around the LP turbine exhaust. This can be avoided by adding a blank piece of duct between the LP hood and condenser (see Fig 3-5).

Because of schedule and space constraints, the condenser assembly may take place adjacent to the TG foundation and then be rolled into position. There are a few methods to roll the condenser into place. One method is to construct a working pad next to the TG foundation and utilize a formed or cast path to guide the condenser into place. Heavy equipment rollers are utilized to roll the equipment, and mechanical means to pull or push the equipment into place are used. To allow for the rolling of the condenser into place, the condenser anchor bolts should utilize couplers for removal during the installation process. Some condenser configurations require a middle support pier, prohibiting the condenser from rolling into place. This pier can be constructed once the condenser is in its design position. One option is to leave embedded plates in the basemat, and construct a steel middle support from these plates afterward.

Fig. 11-6. Condenser installation utilizing air winches and heavy equipment rollers (1)
Source: Courtesy of Kiewit Engineering Group; reproduced with permission.

Fig. 11-7. Condenser installation utilizing air winches and heavy equipment rollers (2)
Source: Courtesy of Kiewit Engineering Group; reproduced with permission.

11.9 ADJUSTABLE VERTICAL SUPPORT DEVICES

Adjustable vertical support devices are precision lifting devices that replace shim packs for aligning equipment. These devices include vertical alignment devices, jacking bolts, and adjusting bolts.

11.9.1 Vertical Alignment Devices

ACI 351.3R, Chapter 5, discusses the use of shims, wedges, or bolts as the interface system between the TG foundation and machine bolts. One interface system not discussed is vertical alignment devices, specialty anchoring, and aligning products available to aid in the installation of TG equipment. These products are a leveling wedge with a spherical washer on top to aid in eliminating any equipment issues that may occur with shims or traditional leveling wedges. One example of vertical alignment devices that is widely used is Fixators (Fig. 11-8). They can be used for lifting as well as alignment, unlike a traditional wedge.

Turbine manufacturers often design large soleplates that are practically difficult to grout properly. What seems to help overcome the poor placement across large machine soleplates is the practice of setting soleplates on steel Fixators, which transfer the machine weights to the concrete foundation, while the grout is simply poured around and underneath them after the initial alignment is achieved.

Vertical alignment devices can be arranged in grout pockets under major equipment baseplates and soleplates. Fig. 11-9 shows a layout of Fixators for supporting an equipment base. Figs. 11-10 and 11-11 show multiple

(a) Fixator w/ Big Adjustment Bolt Enclosure

(b) Fixator w/ Big Adjustment Bolt Enclosure and Enclosed Adjustment Extension

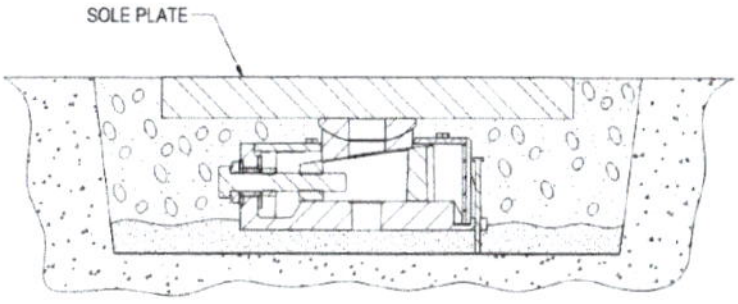

(c) Typical Fixator Fast Track Installation

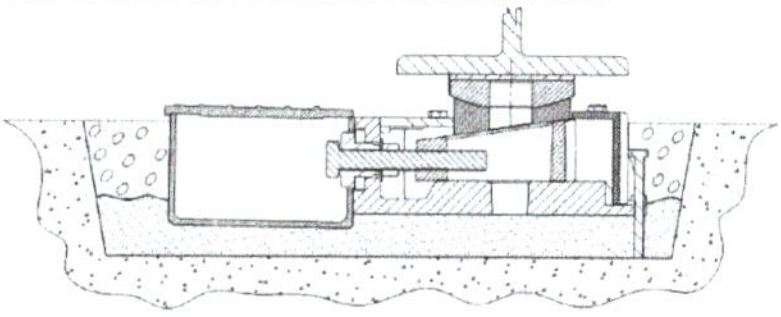

(d) Typical Fixator Direct Installation

Fig. 11-8. Fixator examples
Source: Courtesy of UNISORB; reproduced with permission.

Fig. 11-9. Vertical alignment device (Fixators) layout for equipment base support
Source: Courtesy of Kiewit Engineering Group; reproduced with permission.

Fig. 11-10. Vertical alignment devices set up and grouted into place
Source: Courtesy of Robt. L. Rowan & Assoc., Inc.; reproduced with permission.

Fig. 11-11. Baseplate/soleplate placed on top of vertical alignment devices
Source: Courtesy of Robt. L. Rowan & Assoc., Inc.; reproduced with permission.

vertical alignment devices and anchor bolts positioned at the top of a concrete pedestal, followed by soleplates placement and grout pour.

Also see Section 9.4 for more discussion on grouting.

11.9.2 Jacking Bolt

A jacking bolt, as shown in Fig. 11-12, is a stud that is threaded into and through the equipment soleplate and projects down into the grout pocket. The purpose of this hardware is to help adjust the levelness of the soleplates prior to grout. When jacking bolts are used, more than one is required with a minimum of three being best.

11.9.3 Adjusting Bolts

Adjusting bolts, also referred to as leveling screws, are used to set and align soleplates. During the alignment process, the elevation and level of the foundation soleplate is set by the use of adjusting bolts. A typical configuration will have the adjusting bolts threaded into a nut, which is tack-welded to the side of the foundation baseplate and grouted into place (Fig. 11-13). Adjusting bolts are used only to level the baseplate or soleplate prior to setting the major equipment, since they are not designed to support the full weight of the equipment.

11.10 STAY-IN-PLACE FORMWORK

There are benefits to investigating the use of stay-in-place formwork or modular formwork support systems (ASCE 2014). Construction work at

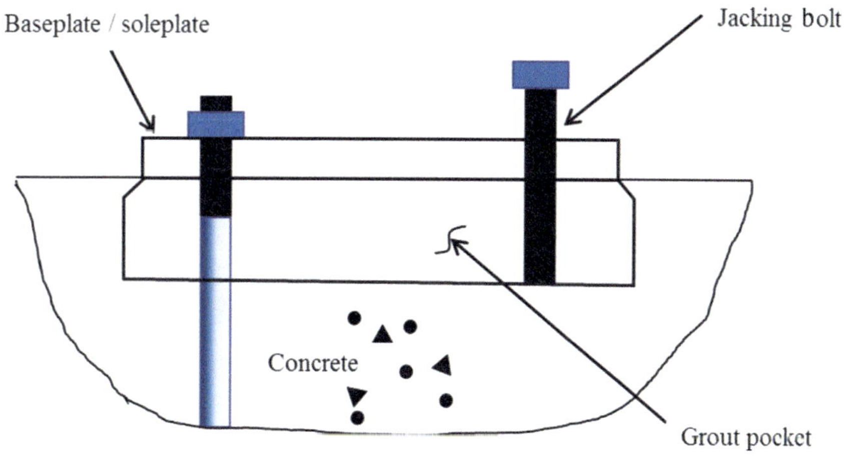

Fig. 11-12. Jacking bolt example

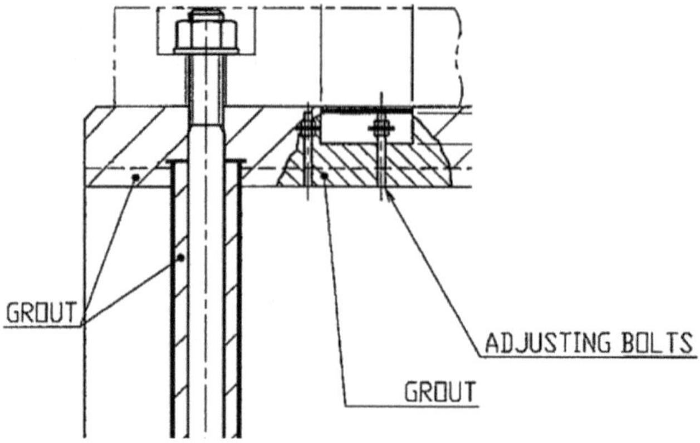

Fig. 11-13. Adjusting bolts

heights can be reduced by building modules at grade. Eliminating the need for traditional formwork and shoring provides opportunities in optimizing construction sequences for other project activities.

REFERENCES

ACI (American Concrete Institute). (2007). "Report on thermal and volume change effects on cracking of mass concrete." Reported by ACI Committee 207, *ACI 207.2R-07*, Detroit.

ACI. (2010). "Specifications for structural concrete." Reported by ACI Committee 301, *ACI 301-10*, Detroit.

ACI. (2004. Reapproved in 2011). "Foundations for dynamic equipment," Reported by ACI Committee 351, *ACI 351.3R-04*, Detroit.

ACI. (2014). "Building code requirements for reinforced concrete." *ACI 318-14*, Detroit.

ASCE. (2014). "Design loads on structure during construction." *ASCE 37-14*, Reston, VA.

Casey, M. J., and Urgessa, G. S. (2013). *Rebar cage construction and safety: best practices*. ASCE, Reston, VA.

Ryan, J. L., Gidwani, N., and Moreschi, L. M. (2012). "Modular stay-in-place formwork for improved construction performance." *Structure* (Sep), 31–33. Reston, VA.

APPENDIX A
DYNAMIC IMPEDANCE OF SOIL-SUPPORTED RIGID FOUNDATIONS

This appendix summarizes some of the solutions available in the literature for calculating the dynamic impedance of soil supported foundations. These solutions are applicable to rigid foundations on homogeneous (Section A.1) and layered soils (Section A.2). They can also be used for preliminary analysis of flexible foundations, such as those typically used for STG and CTG machines. Additionally, the dynamic impedances discussed here are helpful for gaining physical insight into the behavior of soil-supported foundations under dynamic loads.

It must be noted that in practice it is common to cap the damping and use a wide range of soil properties to justify the application of the simple formulas discussed in this appendix for modeling the dynamic response of flexible foundations. In many cases, this practical approach yields conservative results. However, it is not bulletproof, given the complex behavior of flexible foundations, especially those supported on layered soils (e.g., NIST 2012). Therefore, in this book, the recommended approach is to use specialized computer codes that properly model and expedite the dynamic analysis of flexible foundations supported on soil.

A.1 DYNAMIC IMPEDANCE FORMULAS FOR RIGID FOUNDATIONS ON HOMOGENEOUS SOIL

The tables and figures included in this section are reported by Gazetas (1991). The formulas for calculating the frequency dependent soil "spring" and "dashpot" coefficients are presented in Table A-2 and Table A-3. Fig. A-1 sketches the soil-foundation systems covered by each table. Specifically: (a) Table A-2 and the accompanying set of graphs refer to foundations

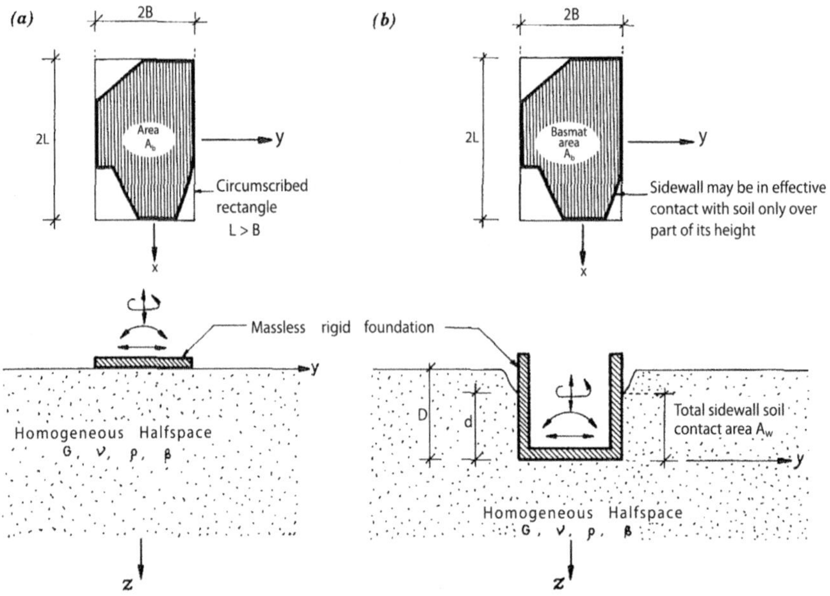

Fig. A-1. Soil-foundation systems considered in this section; (a) surface foundation of arbitrary shape (see Table A-2 and Fig. A-2); (b) embedded foundation of arbitrary basemat shape (see Table A-3 and Fig. A-3)
Source: Gazetas 1991.

Table A-1. Geometrical Properties of a Rectangular Footing Having Dimensions 2B and 2L, as Shown in Fig. A-1

Area	Area moment of inertia	Area polar moment of inertia
$A_b = 4BL$	$I_{bx} = \dfrac{4LB^3}{3}$ and $I_{by} = \dfrac{4BL^3}{3}$	$J_b = \dfrac{4BL(B^2 + L^2)}{3}$

of any solid basemat shape resting on the surface of a homogeneous half-space; and (b) Table A-3 and the related graphs are for foundations with any solid basemat shape partially or fully embedded in a homogeneous half-space. The following steps can be used for calculating the dynamic impedance of surface foundations using the subject tables and figures.

1. Verify that the soil profile is, or can be approximated as, homogeneous and obtain the best estimate (BE) of the following soil properties:

   G_{max}: low strain shear modulus
   ν: Poisson's ratio

ρ: Total soil density (solid + fluid phase)
β_0: Low strain damping ratio (for most soil ranges between 2% and 3%)

$V_s = \sqrt{G_{max}/\rho}$: shear wave velocity

$V_p = V_s\sqrt{\dfrac{2(1-\nu)}{(1-2\nu)}}$: compressional wave velocity

$V_{La} = \dfrac{3.4V_s}{\pi(1-\nu)}$: Lysmer's analog wave velocity

2. Calculate the lower bound (LB) and upper bound (UB) values of G_{max}, V_s, V_p, and V_{La}.
3. Calculate the area (A_b), the centroid axes x and y, the area moments of inertia about x and y (I_{bx}, I_{by}), and the polar moment of inertia about z (J_b) of the actual contact surface between the foundation and subgrade, as detailed in Table A-1 for a rectangular foundation.

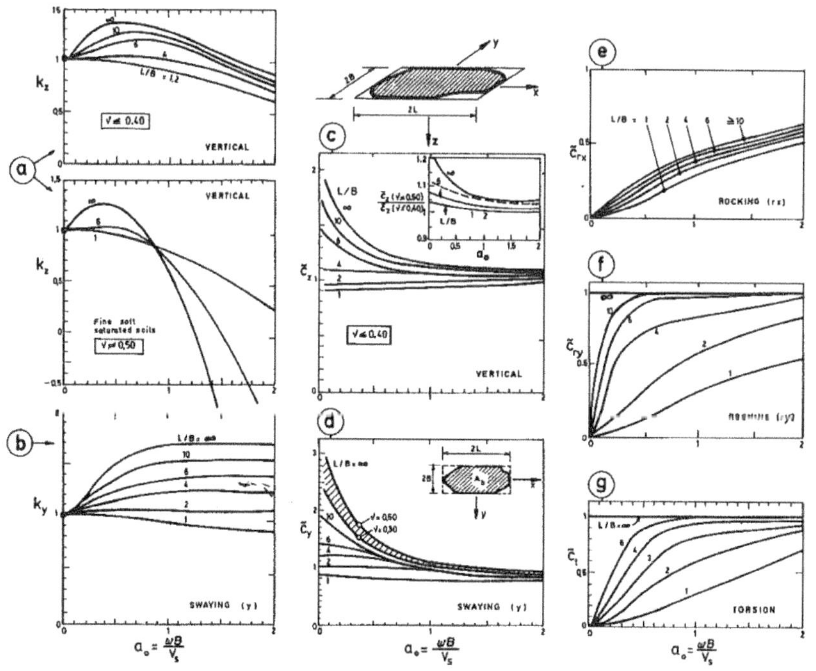

Fig. A-2. Dimensionless graphs for determining dynamic stiffness and damping coefficients of surface foundations; graphs accompanying Table A-2
Source: Gazetas 1991.

Table A-2. Dynamic Stiffness and Dashpot Coefficients for Arbitrarily Shaped Foundations on the Surface of a Homogeneous Half-space

	Dynamic stiffness $\overline{K} = K \times k(a_0)$			
	Static stiffness K			
Vibration mode	General shape (foundation-soil contact surface is of area A_b and has a circumscribed rectangle $2L$ by $2B$; $L > B$)	Square $L = B$	Dynamic stiffness coefficient $k(a_0)$ (general shape; $0 \le a_0 \le 2$)	Radiation dashpot coefficient C (general shapes)
Vertical, z	$K_z = \frac{2GL}{1-v}(0.73 + 1.54\chi^{0.75})$ with $\chi = \frac{A_b}{4L^2}$	$K_z = \frac{4.54GB}{1-v}$	$k_z = k_z\left(\frac{L}{B}, v; a_0\right)$ is plotted in Graph "a" (Fig. A-2)	$C_z = (\rho V_{La} A_b)\overline{c}_z$ $\overline{c}_z = \overline{c}_z(L/B, v; a_0)$ is plotted in Graph "c" (Fig. A-2)
Horizontal, y (in lateral direction)	$K_y = \frac{2GL}{2-v}(2 + 2.50\chi^{0.85})$	$K_y = \frac{9GB}{2-v}$	$k_y = k_y\left(\frac{L}{B}; a_0\right)$ is plotted in Graph "b" (Fig. A-2)	$C_y = (\rho V_s A_b)\overline{c}_y$ $\overline{c}_y = \overline{c}_y(L/B; a_0)$ is plotted in Graph "d" (Fig. A-2)
Horizontal, x (in longitudinal direction)	$K_x = K_y - \frac{0.2}{0.75-v}GL\left(1 - \frac{B}{L}\right)$	$K_x = K_y$	$k_x \approx 1$	$C_x \approx \rho V_s A_b$

Rocking, *rx* (around longitudinal x-axis)	$K_{rx} = \frac{G}{1-v} I_{bx}^{0.75} \left(\frac{L}{B}\right)^{0.25} \left(2.4 + 0.5\frac{B}{L}\right)$ I_{bx} (I_{by}) is area moment of inertia of the foundation-soil contact surface around the $x(y)$-axis.	$K_{rx} = \frac{3.6GB^3}{1-v}$	$k_{rx} \approx 1 - 0.20a_o$	$C_{rx} = (\rho V_{La} I_{bx}) \cdot \bar{c}_{rx}$ $\bar{c}_{rx} = \bar{c}_{rx}(L/B;\ a_0)$ is plotted in Graphs "*e*" and "*f*" (Fig. A-2)
Rocking, *ry* (around lateral axis)	$K_{ry} = \frac{G}{1-v} I_{by}^{0.75} \left[3\left(\frac{L}{B}\right)^{0.15}\right]$	$K_{ry} = K_{rx}$	$\begin{cases} v < 0.45: \\ k_{ry} \approx 1 - 0.30a_0 \\ v \approx 0.50: \\ k_{ry} \approx 1 - 0.25a_0 \left(\frac{L}{B}\right)^{0.30} \end{cases}$	$C_{ry} = (\rho V_{La} I_{by}) \cdot \bar{c}_{ry}$ $\bar{c}_{ry} = \bar{c}_{ry}(L/B; a_0)$ is plotted in Graph "*g*" (Fig. A-2)
Torsional	$K_t = GJ_b^{0.75}\left[4 + 11\left(1 - \frac{B}{L}\right)^{10}\right]$ with $J_b = I_{bx} + I_{by}$ being the polar moment of the soil-foundation contact surface	$k_t = 8.3GB^3$	$k_t \approx 1 - 0.14a_0$	$C_t = (\rho V_s J_b) \cdot \bar{c}_t$ $\bar{c}_t = \bar{c}_t(L/B; a_0)$

Notes: Refer to accompanying Fig. A-2.

$a_0 = \frac{\Omega B}{V_s};\ V_{La} = \frac{3.4V_s}{\pi(1-v)}$

Source: Gazetas 1991.

4. Obtain the equivalent foundation dimensions, as shown in Fig. A-1, in other words, the dimensions B and L of the rectangle circumscribing the contact surface between the foundation and subgrade.
5. Calculate the nondimensional frequency $a_0 = \omega B/V_s$ corresponding to the operating frequency of the machine ($\omega = \Omega_o$) and other frequencies of interest ($\omega = \Omega$) taking place during the machine startup and shutdown.
6. Use Table A-2 for calculating the static stiffness K_j corresponding to each vibration mode.
7. Use Table A-2 and Fig. A-2 for calculating the dynamic stiffness $\overline{K}_j$ and the dashpot coefficient C_j, for each frequency and vibration mode of interest.
8. If applicable, use Table A-3 and Fig. A-3 to account for embedment effects.

A.2 IMPEDANCE FUNCTIONS OF RIGID FOUNDATIONS USING THE CONE MODEL APPROACH

The steps required for the calculation of the dynamic impedance of rigid foundations can be easily implemented and understood using cone models. This approach allows simplifying the solution of complex 3D elastodynamics problems, given that only one-dimensional wave propagation is accounted for. Furthermore, the foundation is assumed to be circular and rigid; while the soil is modeled as layered linear elastic media, as shown in Fig. 5-9 and Fig. A-4. As a consequence, only approximated solutions can be obtained. Nevertheless, such solutions are generally within 20% of those obtained from more rigorous and complicated approaches using elastodynamics theory, as shown by Wolf and Deeks (2004). However, this loss of precision is compensated by the ease of implementation and understanding of the method, given that only familiar strength of material concepts are used for its formulation. Furthermore, subroutines based on the cone approach can be coded using the algorithms provided by Wolf and Deeks (2004). Typical results of the method, for the foundation shown in Fig. A-4, are plotted and compared against the "exact" solution in Fig. A-5. As can be seen, engineering accuracy, within 20% of the "exact elastodynamics" solution, is achieved for all the vibration modes throughout the complete frequency range.

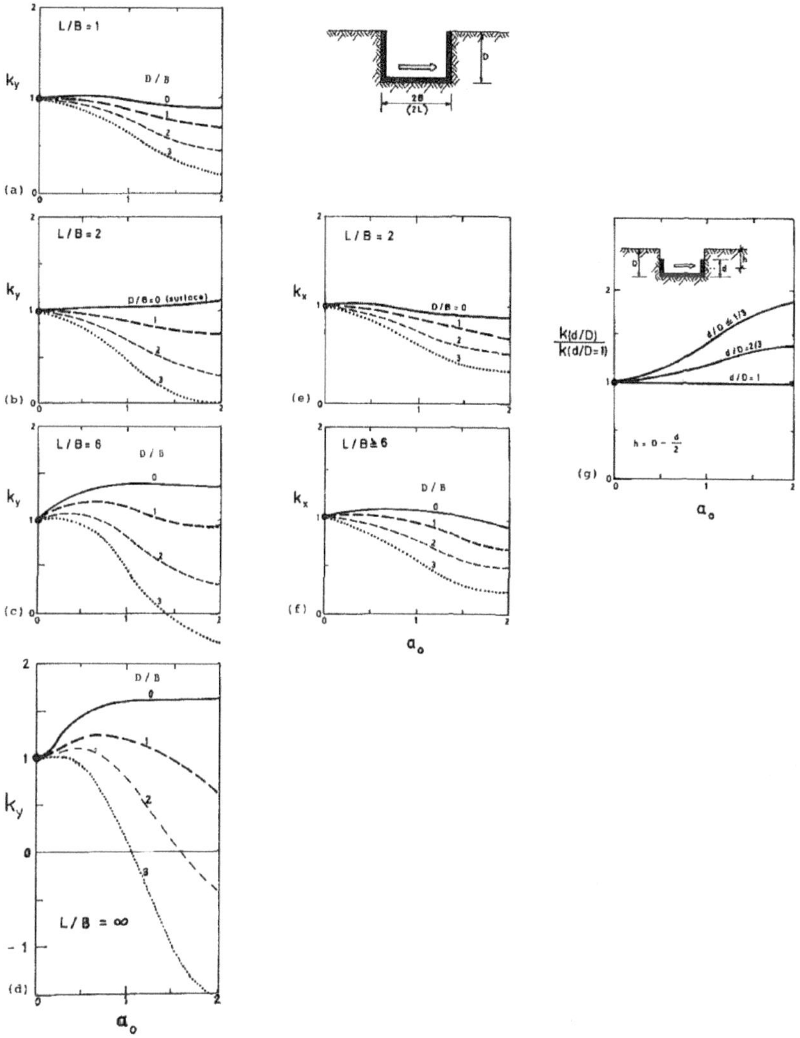

Fig. A-3. Dimensionless graphs for determining dynamic stiffness coefficients of fully and partially embedded foundations; graphs accompanying Table A-3 Source: Gazetas 1991.

Table A-3. Dynamic Stiffness and Dashpot Coefficients for Arbitrarily Shaped Foundations Partially or Fully Embedded in a Homogeneous Half-space

	Dynamic stiffness $\bar{K}_{emb} = K_{emb} \times k_{emb}(a_0)$		Radiation dashpot coefficient $C_{emb}(a_0)$	
Vibration mode	Static stiffness K_{emb} For foundations with arbitrarily shaped basement A_b of circumscribed rectangle $2L$ by $2B$; total sidewall soil contact area A_w (or contact sidewall soil contact height d)	Dynamic stiffness coefficient $k_{emb}(a_0)$ $0 < a_0 \le 2$	General foundation shape	Rectangular foundation $2L$ by $2B$ by d
Vertical, z	$K_{z.emb} = K_{z.sur}\left[1+\frac{1}{21}\frac{D}{B}(1+1.3\chi)\right] \times \left[1+0.2\left(\frac{A_w}{A_b}\right)^{2/3}\right]$ $K_{z.sur}$ obtained from Table A-1 A_w = actual sidewall soil contact area; for constant effective contact height d along the perimeter $A_w = (d)(Perimeter)$ $\chi = A_b/4L^2$	$v \le 0.40$ Fully embedded: $k_{z.emb} = k_{z.sur}\left[1-0.09\left(\frac{D}{B}\right)^{3/4} a_0^2\right]$ In a trench: $k_{z.tre} = k_{z.sur}\left[1+0.09\left(\frac{D}{B}\right)^{3/4} a_0^2\right]$ Partially embedded: estimate by interpolating between the two $v \approx 0.50$ Fully embedded, $L/B = 1-2$ $k_{z.emb} = 1-0.09\left(\frac{D}{B}\right)^{3/4} a_0^2$ Fully embedded, $L/B \ge 3$ $k_{z.emb} = 1-0.35\left(\frac{D}{B}\right)^{3/4} a_0^{3.5}$	$C_{z.emb} = C_{z.sur} + \rho V_s A_w$ with $C_{z.sur}$ and C_z according to Table A-1	$C_{z.emb} = 4\rho V_{La} BL\bar{c}_z + 4\rho V_s (B+L)d$

Horizontal y or x	$K_{y.emb} = K_{y.sur}(1+0.15\sqrt{D/3}) \times \left[1+0.52\left(\frac{h}{B}\frac{A_w}{L^2}\right)^{0.4}\right]$ $K_{y.sur}$ obtained from Table A-1 $K_{x.emb}$ similarly computed from $K_{x.sur}$	$k_{y.emb}$ and $k_{x.emb}$ can be estimated in terms of L/B, D/B, and d/B for each value of a_0 from the graphs accompanying this table (Fig. A-3).	$C_{y.emb} = C_{y.sur} + \rho V_s A_{ws} + \rho V_{La} A_{wce}$ $A_{ws} = \Sigma(A_{wi}\sin\theta_i) =$ total effective sidewall area shearing the soil $A_{wce} = \Sigma(A_{wi}\cos\theta_i) =$ total effective sidewall area compressing the soil θ_i = angle of inclination of surface A_{wi} from loading direction $C_{y.sur}$ obtained from Table A-1 $C_{x.emb}$ similarly computed from $C_{x.sur}$	$C_{y.emb} = 4\rho V_s BL\bar{c}_y + 4\rho V_s Bd + 4\rho V_{La} Ld$ $\bar{c}_y$ according to Table A-1
Rocking, rx (around longitudinal axis)	Expressions valid for any basemat shape but constant effective contact height d along the perimeter $K_{rx.emb} = K_{rx.sur} \times \left\{1+1.26\frac{d}{B}\left[1+\frac{d}{B}\left(\frac{d}{D}\right)^{-0.2}\sqrt{\frac{B}{L}}\right]\right\}$	$k_{rx.emb} = k_{rx.sur}$	$C_{rx.emb} = C_{rx.sur} + \rho V_{Lo} I_{wce}\bar{c}_1 + \rho V_s\left(J_{ws} + \sum_i [A_{wcei}\Delta_i^2]\right)\bar{c}_1$ $\bar{c}_1 = 0.25 + 0.65\sqrt{a_0}\left(\frac{d}{D}\right)^{-a0/2}\left(\frac{D}{B}\right)^{-1/4}$ I_{wce} = total moment of inertia about their base axes parallel to x of all sidewall surfaces effectively compressing the soil	$C_{rx.emb} = \frac{4}{3}\rho V_{La} B^3 L\bar{c}_{rx} + \frac{4}{3}\rho V_{La} d^3 L\bar{c}_1 + \frac{4}{3}\rho V_s Bd(B^2+d^2)\bar{c}_1 + 4\rho V_s B^2 dL\bar{c}_1$ with $\bar{c}_1$ as in the preceding column and $\bar{c}_{rx}$ according to Table A-1
Rocking, ry (around lateral axis)	$K_{ry.emb} = K_{ry.sur} \times \left\{1+0.92\left(\frac{d}{L}\right)^{0.6}\left[1.5+\left(\frac{d}{L}\right)^{1.9}\left(\frac{d}{D}\right)^{-0.6}\right]\right\}$	$k_{ry.emb} = k_{ry.sur}$	Δ_i = distance of surface A_{wcei} from the x axis J_{ws} = polar moment of inertia about their base axes parallel to x of all sidewall surfaces effectively shearing the soil $C_{ry.emb}$ is similarly evaluated from $C_{ry.sur}$ with y replacing x and in the equation for $\bar{c}_1$, L replacing B	

Table A-3. (Continued)

Vibration mode	Dynamic stiffness $\overline{K}_{emb} = K_{emb} \times k_{emb}(a_0)$ — Static stiffness K_{emb} For foundations with arbitrarily shaped basement A_b of circumscribed rectangle $2L$ by $2B$; total sidewall soil contact area A_w (or contact sidewall soil contact height d)	Dynamic stiffness coefficient $k_{emb}(a_0)$ $0 < a_0 \le 2$	Radiation dashpot coefficient $C_{emb}(a_0)$ — General foundation shape	Rectangular foundation $2L$ by $2B$ by d
Coupling term swaying-rocking (x, ry)	$K_{xry.emb} \approx \frac{1}{3} d K_{x.emb}$	$k_{rxy.emb} \approx k_{yrx.emb} \approx 1$	$C_{xry.emb} \approx \frac{1}{3} d C_{x.emb}$	As in the previous column
swaying-rocking (y, rx)	$K_{yrx.emb} \approx \frac{1}{3} d K_{y.emb}$		$C_{yrx.emb} \approx \frac{1}{3} d C_{y.emb}$	
Torsional	$K_{t.emb} = K_{t.sur} \times \left[1 + 1.4\left(1 + \frac{B}{L}\right)\left(\frac{d}{B}\right)^{0.8}\right]$	$k_{t.emb} \approx k_{t.sur}$	$C_{t.emb} = C_{t.sur} + \rho V_{La} J_{wce} \overline{c}_2 + \rho V_s \sum \left[A_{wi} \cdot \Delta^2_{zi}\right] \cdot \overline{c}_2$ $\overline{c}_2 \approx \left(\frac{d}{D}\right)^{-0.5} \cdot \frac{a_0^2}{a_0^2 + \frac{1}{2}(L/B)^{-1.5}}$ J_{wce} = total moment of inertia of all sidewall surfaces effectively compressing the soil about the projection of the z axis onto their plane Δ_{zi} = distance of surface A_{wi} from the z axis	$C_{t.emb} = \frac{4}{3} \rho V_s BL(B^2 + L^2)\overline{c}_t + \frac{4}{3} \rho V_{La} d(L^3 + B^3)\overline{c}_2 + 4 \rho V_s dBL(B + L)\overline{c}_2$ With $\overline{c}_2$ as in the preceding column and $\overline{c}_t$ according to Table A-1

Notes: Refer to accompanying Fig. A-3.
Source: Gazetas 1991.

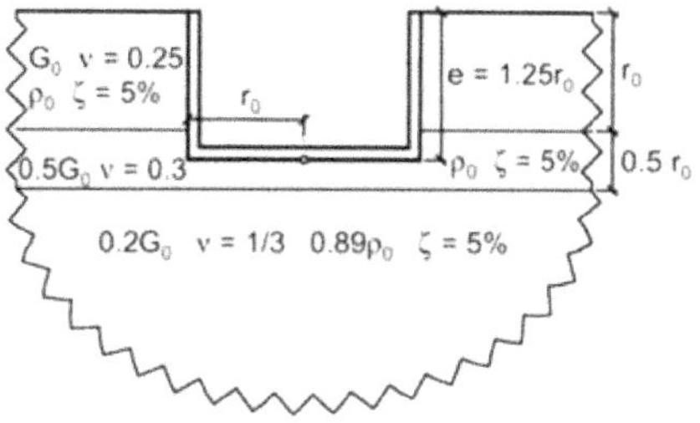

Fig. A-4. Cylindrical foundation embedded in site consisting of two layers overlying flexible half-space
Source: Wolf and Deeks (2004); reproduced with permission from Elsevier.

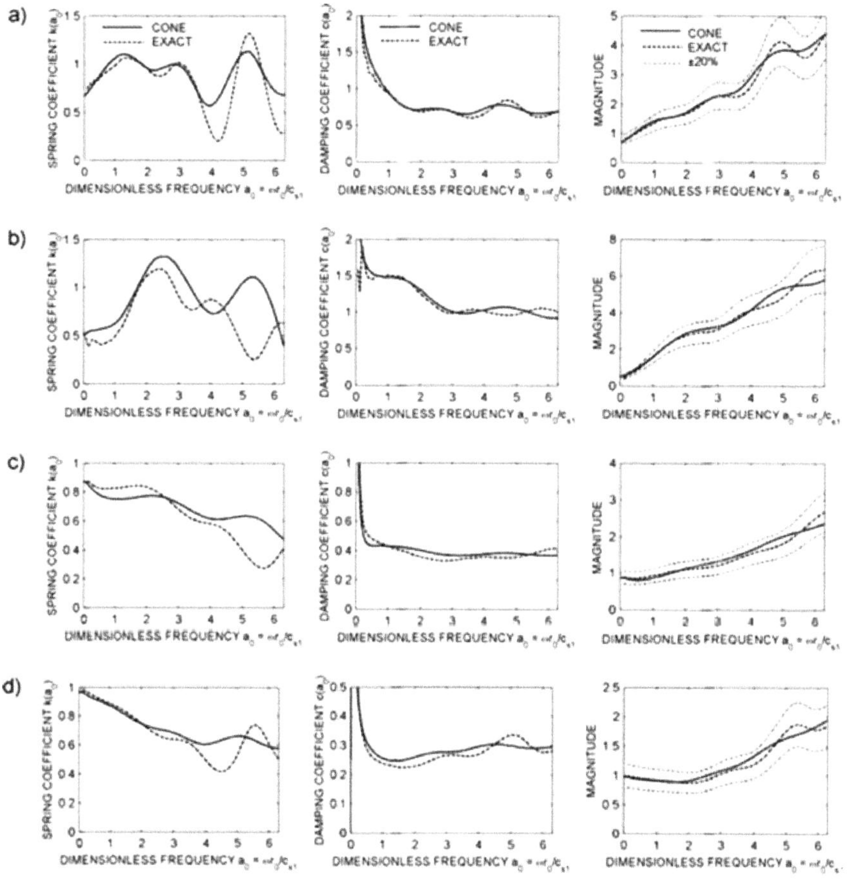

Fig. A-5. Dynamic stiffness coefficients of disk on two layers overlying flexible half-space; (a) horizontal; (b) vertical; (c) rocking; (d) torsional
Source: Wolf and Deeks (2004); reproduced with permission from Elsevier.

REFERENCES

Gazetas, G. (1991). "Formulas and charts for impedances of surface and embedded foundations." *J. Geotech. Engrg. Div.* 117(9), 1363–1381.

NIST (National Institute of Standards and Technology). (2012). *Soil-structure interaction for building structures*. GCR 12-917-21 U.S. Department of Commerce, Washington, DC.

Wolf, J. P., and Deeks, A. J. (2004). *Foundation vibration analysis: A strength-of-materials approach*. Elsevier, Amsterdam.

APPENDIX B
DYNAMIC IMPEDANCE OF PILE-SUPPORTED FOUNDATIONS

This appendix summarizes some of the solutions available in the literature for calculating the dynamic impedance of pile-supported foundations. These solutions are applicable to single piles embedded in uniform or parabolic soil profiles, and to small groups of piles in uniform soil. In any case, the subject solutions are very helpful for gaining physical insight into the behavior of pile-supported foundations under dynamic loads.

It must be noted that in practice it is common to cap the damping and use a wide range of soil properties to justify the use of the simple solutions discussed herein, for the analysis of pile foundations in layered soils and/or for large pile groups. In many cases, this practical approach yields conservative results. However, it is not recommended for piles embedded in soil layers with strongly varying and contrasting soil properties, or for foundations on large pile groups, which may exhibit complex behavior. Therefore, in this book, the recommended approach is to use specialized computer codes that properly model and expedite the dynamic analysis of pile-supported foundations.

B.1 IMPEDANCE FUNCTIONS OF SINGLE PILES ON HOMOGENEOUS AND PARABOLIC SOIL PROFILES

For the purposes of this appendix the dynamic impedance of a single pile is represented mathematically as

$$K_j(a_0) = k_j(a_0) + i\Omega c_j(a_0) \tag{B-1}$$

Table B-1. Stiffness and Damping Constants for Single Piles

Vertical	Horizontal	Rocking	Coupling	Torsion
$k_z = \frac{E_p A}{R} f_{v1}$	$k_x = \frac{E_p I}{R^3} f_{u1}$	$k_{ry} = \frac{E_p I}{R} f_{\varphi 1}$	$k_{xry} = \frac{E_p I}{R^2} f_{c1}$	$k_{rz} = \frac{G_p J}{V_s} f_{\eta 1}$
$c_z = \frac{E_p A}{V_s} f_{v2}$	$c_x = \frac{E_p I}{R^2 V_s} f_{u2}$	$c_{ry} = \frac{E_p I}{V_s} f_{\varphi 2}$	$c_{xry} = \frac{E_p I}{R V_s} f_{c2}$	$c_{rz} = \frac{G_p J}{V_s} f_{\eta 2}$

In particular, the dynamic stiffness, k_j, and the equivalent viscous damping, c_j, coefficients for individual motions of the pile head suggested by Novak (1974) are shown in Table B-1. These coefficients are a function of the pile's elastic modulus, E_p, cross-sectional area, A, and moment of inertia and torsional stiffness I and G_pJ, respectively. R is the radius of circular piles or equivalent radius for noncircular piles. This radius is related to the soil reactions; therefore, for noncircular piles, it must be calculated from the external pile dimensions such as the width and depth, whereas other pile geometry parameters (e.g., A, I, and J) are computed for the actual pile cross section. The symbol $f_{1,2}$ in Table B-1 represents dimensionless stiffness and damping functions, whose subscript 1 indicates stiffness and 2 indicates damping. These functions depend on the following parameters: the relative stiffness of the pile and soil, E_p/G; the dimensionless frequency, a_0; the slenderness ratio, L/R, in which L = pile length; the material damping (C) of both the soil and pile; the mass ratio ρ/ρ_p, where ρ and ρ_p are the soil and pile mass densities; the variation of soil and pile properties with depth; and the tip and head conditions. However, E_p/G, the soil profile and, for the vertical direction, the tip condition have the strongest effect on the stiffness. Table B-2 lists the horizontal stiffness and damping functions ($f_{1,2}$) for a few particular cases, where the superscript p applies to pin-headed piles. Similarly, Fig. B-1 shows the vertical stiffness and damping functions ($f_{v1,2}$). Table B-2 and Fig. B-1 functions are calculated with material damping ratio D = 0.025 for soil and 0.005 for piles, mass ratio $\rho/\rho_p = 1$, and $a_0 = 0.3$. For other cases, please refer to Novak and El Sharnouby (1983) and NIST (2012).

It must be noted that damping functions $f_{i,2}$ reported in Table B-2 and Fig. B-1 are only valid for excitation frequencies greater than the fundamental frequency of the soil layer (i.e., ω_{su} and ω_{sv} for the horizontal and vertical directions). Therefore, damping corrections are required for low excitation frequencies ($\Omega < \omega_{si}$), as described by Novak and El Sharnouby (1983). In particular, for a homogeneous layer with soil shear wave velocity V_s and thickness H overlying a rigid stratum, these fundamental frequencies are given by Eq. (B-2).

Horizontal direction:

$$\omega_{su} = \frac{\pi V_s}{2H} \tag{B-2a}$$

Vertical direction:

$$\omega_{sv} = \frac{\pi V_s}{2H}\sqrt{\frac{2(1-v)}{1-2v}} \tag{B-2b}$$

Table B-2. Stiffness and Damping Parameters for Horizontal Response of Piles in Homogeneous and Parabolic Soil Profiles ($L/R > 25$ for homogeneous soil and $L/R > 30$ for parabolic soil profile)

		Stiffness parameters				Damping parameters			
ν	E_p/G	$f_{\psi 1}$	f_{c1}	f_{u1}	f^p_{u1}	$f_{\psi 2}$	f_{c2}	f_{u2}	f^p_{u2}
Homogeneous soil profile									
0.25	10,000	0.2135	−0.0217	0.0042	0.0021	0.1577	−0.0333	0.0107	0.0054
	2,500	0.2998	−0.0429	0.0119	0.0061	0.2152	−0.0646	0.0297	0.0154
	1,000	0.3741	−0.0668	0.0236	0.0123	0.2598	−0.0985	0.0579	0.0306
	500	0.4411	−0.0929	0.0395	0.0210	0.2953	−0.1337	0.0953	0.0514
	250	0.5186	−0.1281	0.0659	0.0358	0.3299	−0.1786	0.1556	0.0864
0.4	10,000	0.2207	−0.0232	0.0047	0.0024	0.1634	−0.0358	0.0119	0.0060
	2,500	0.3097	−0.0459	0.0132	0.0068	0.2224	−0.0692	0.0329	0.0171
	1,000	0.3860	−0.0714	0.0261	0.0136	0.2677	−0.1052	0.0641	0.0339
	500	0.4547	−0.0991	0.0436	0.0231	0.3034	−0.1425	0.1054	0.0570
	250	0.5336	−0.1365	0.0726	0.0394	0.3377	−0.1896	0.1717	0.0957
Parabolic soil profile									
0.25	10,000	0.1800	−0.0144	0.0019	0.0008	0.1450	−0.0252	0.0060	0.0028
	2,500	0.2452	−0.0267	0.0047	0.0020	0.2025	−0.0484	0.0159	0.0076
	1,000	0.3000	−0.0400	0.0086	0.0037	0.2499	−0.0737	0.0303	0.0147
	500	0.3489	−0.0543	0.0136	0.0059	0.2910	−0.1008	0.0491	0.0241
	250	0.4049	−0.0734	0.0215	0.0094	0.3361	−0.1370	0.0793	0.0398
0.4	10,000	0.1857	−0.0153	0.0020	0.0009	0.1508	−0.0271	0.0067	0.0031
	2,500	0.2529	−0.0284	0.0051	0.0022	0.2101	−0.0519	0.0177	0.0084
	1,000	0.3094	−0.0426	0.0094	0.0041	0.2589	−0.0790	0.0336	0.0163
	500	0.3596	−0.0577	0.0149	0.0065	0.3009	−0.1079	0.0544	0.0269
	250	0.4170	−0.0780	0.0236	0.0103	0.3468	−0.1461	0.0880	0.0443

Source: Table 1 in Novak and El Sharnouby (1983).

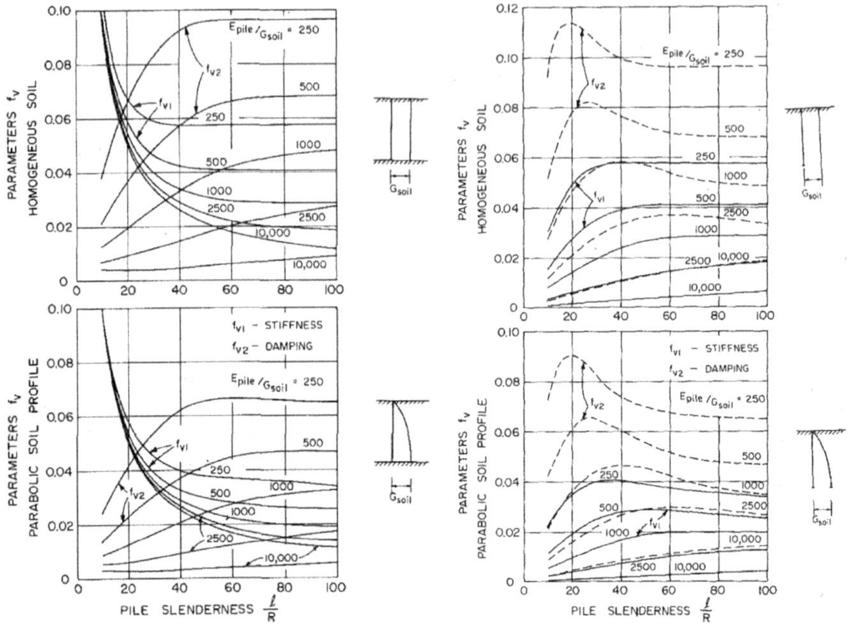

Fig. B-1. Stiffness and damping parameters f_{v1} and f_{v2} for vertical response of end bearing and floating piles
Source: Figs. 4 and 5 in Novak and El Sharnouby (1983).

B.2 DYNAMIC IMPEDANCE OF PILE GROUPS

The behavior of a pile group depends on the distance between individual piles and the rigidity of the pile cap. When the distance between individual piles is large—20 diameters or more—the piles do not affect each other, and the group stiffness and damping are calculated as the summation of the contributions from each individual pile. However, for typical pile spacing, pile-soil-pile interaction (group effect) significantly influences the stiffness and damping of the group and cannot be ignored. The following sections address the calculation of pile group effects using a simplified approach based on dynamic interaction factors.

B.2.1 Dynamic Impedance of Rigid Pile Cap Neglecting Pile to Pile Interaction

The starting point for the analysis of a group of piles connected by a rigid cap is the calculation of the group impedance neglecting pile to pile interaction. Subsequently, the group effect can be accounted for by using frequency dependent group factors as discussed by El et al. (1992) and/or the approach discussed in Section B.2.2.

Neglecting pile to pile interaction, the stiffness and damping of a pile group are determined by summation of the dynamic stiffness and damping of the individual piles with respect to a reference point, the pile cap centroid. In particular, the group stiffness and damping are defined as the forces that must be applied at the pile cap centroid to produce a unit displacement or rotation, as shown in Fig. B-2 for rocking and torsional vibration.

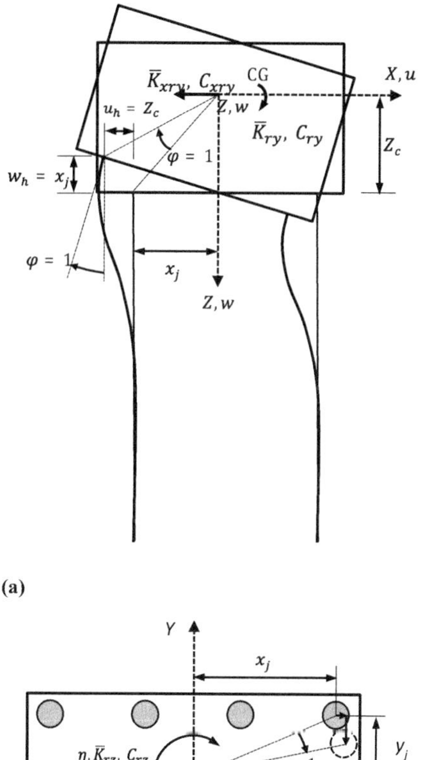

(a)

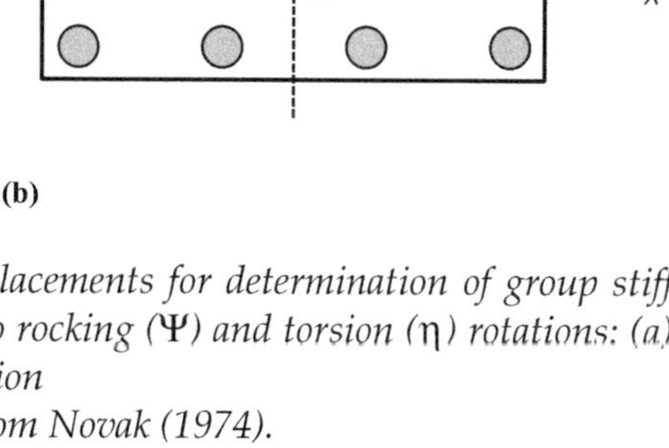

(b)

Fig. B-2. Pile displacements for determination of group stiffness and damping constants related to rocking (Ψ) and torsion (η) rotations: (a) rocking vibration; (b) torsional vibration
Source: Adapted from Novak (1974).

From this definition, the stiffness and damping of the pile group, with respect to the centroid of the pile cap, are provided in Table B-3. Where the pile cap is assumed infinitely rigid, $\bar{K}_j$ and C_j are the dynamic pile group stiffness and damping in the jth vibration mode, N is the number of piles, and other terms are defined in Fig. B-2.

Table B-3. Dynamic Impedances of a Rigid Pile Cap Neglecting Pile to Pile Interaction

Motion	Stiffness	Damping
Vertical	$\bar{K}_z = \sum_{j=1}^{N} k_{zj}$	$C_z = \sum_{j=1}^{N} c_{zj}$
Horizontal	$\bar{K}_x = \sum_{j=1}^{N} k_{xj}$	$C_x = \sum_{j=1}^{N} c_{xj}$
Horizontal	$\bar{K}_y = \sum_{j=1}^{N} k_{yj}$	$C_x = \sum_{j=1}^{N} c_{yj}$
Rocking about y	$\bar{K}_{ry} = \sum_{j=1}^{N} k_{\mathrm{ryj}} + k_{zj} x_j^2 + k_{xj} z_c^2 - 2k_{xryj} z_c$	$C_{ry} = \sum_{j=1}^{N} c_{\mathrm{ryj}} + c_{zj} x_j^2 + c_{xj} z_c^2 - 2c_{\mathrm{xryj}} z_c$
Rocking about x	$\bar{K}_{rx} = \sum_{j=1}^{N} k_{\mathrm{rxj}} + k_{zj} y_j^2 + k_{yj} z_c^2 - 2k_{yrxj} z_c$	$C_{rx} = \sum_{j=1}^{N} c_{\mathrm{rxj}} + c_{zj} y_j^2 + c_{zj} y_j^2 + c_{yj} z_c^2 - 2c_{yrxj} z_c$
Coupling	$\bar{K}_{xry} = \sum_{j=1}^{N} k_{\mathrm{xry}\psi\mathrm{j}} - k_{xj} z_c$	$C_{xry} = \sum_{j=1}^{N} c_{\mathrm{xryj}} - c_{xj} z_c$
Coupling	$\bar{K}_{yrx} = \sum_{j=1}^{N} k_{\mathrm{yrx}\psi\mathrm{j}} - k_{yj} z_c$	$C_{xry} = \sum_{j=1}^{N} c_{\mathrm{yrxj}} - c_{yj} z_c$
Torsion[1]	$\bar{K}_{rz} = \sum_{j=1}^{N} k_{rzj} + k_{xj}(x_j^2 + y_j^2)$	$C_{rz} = \sum_{j=1}^{N} c_{rzj} + c_{xj}(x_j^2 + y_j^2)$

[1]Valid for circular pile cross section (i.e., $k_x = k_y$)

B.2.2 Group Impedance Considering Pile-Soil-Pile Interaction

When piles are in a group, they interact with each other because the displacement of one pile contributes to the displacement of the others. To obtain an accurate analysis of dynamic behavior of pile groups, it is necessary to use a suitable computer program. However, a simplified approximate analysis can be formulated on the basis of interaction factors, α_{ij}, introduced by Poulos (1971) for pile groups under static loads and extended to the dynamic case by Kaynia and Kausel (1982). In any case, these factors are defined considering the physical interaction between two piles (Eq. 5-38), which can be represented mathematically by Eq. (B-3), where u_{ij} is the head displacement of pile i owing to a unit load applied on the head of pile j. Additionally, it must be noted that the respective stiffness of piles 1 and 2 are $k_1 = \frac{1}{u_{11}}$ and $k_2 = \frac{1}{u_{22}}$, and $\alpha_{12} = \alpha_{21}$ furthermore, for identical piles $u_{11} = u_{22}$.

$$A = \begin{bmatrix} \frac{u_{11}}{u_{11}} & \frac{u_{12}}{u_{11}} \\ \frac{u_{21}}{u_{22}} & \frac{u_{22}}{u_{22}} \end{bmatrix} = \begin{bmatrix} 1 & \alpha_{12} \\ \alpha_{21} & 1 \end{bmatrix} \tag{B-3}$$

To illustrate the derivation of a pile group stiffness, consider two piles under a vertical or horizontal load, with their heads rigidly connected. If P_G is the total load and P_1 and P_2 are the loads respectively acting on piles 1 and 2, the pile displacements and forces can be calculated as follows, using the given definitions.

$$P_G = P_1 + P_2 \tag{B-4}$$

$$u_1 = u_{11}P_1 + u_{12}P_2 = u_{11}(P_1 + \alpha_{12}P_2) = \frac{1}{k_1}(P_1 + \alpha_{12}P_2) \tag{B-5}$$

$$u_2 = u_{21}P_1 + u_{22}P_2 = u_{22}(\alpha_{21}P_1 + P_2) = \frac{1}{k_2}(P_1 + \alpha_{21}P_2) \tag{B-6}$$

$$U = K^{-1} AP \quad \therefore \quad P = A^{-1}KU \tag{B-7}$$

$$U = \begin{Bmatrix} u_1 \\ u_2 \end{Bmatrix} P = \begin{Bmatrix} P_1 \\ P_2 \end{Bmatrix} P_G = \begin{Bmatrix} 1 & 1 \end{Bmatrix} \begin{Bmatrix} P_1 \\ P_2 \end{Bmatrix} = I^T P \; K = \begin{bmatrix} k_1 & 0 \\ 0 & k_2 \end{bmatrix}$$

The group stiffness can be easily calculated using these equations and taking into account that for rigidly connected piles $u_1 = u_1 = u_G$; therefore

$$U = \begin{Bmatrix} u_1 \\ u_2 \end{Bmatrix} = \begin{Bmatrix} 1 \\ 1 \end{Bmatrix} u_G = I u_G \tag{B-8}$$

$$P_G = I^T P = I^T A^{-1} K I u_G \tag{B-9}$$

$$P_G = K_G u_G \quad \therefore \quad K_G = I^T A^{-1} K I \tag{B-10}$$

Eq. (B-10) is easily generalized for a group of n piles, the only difference being that all matrices are of order n by n, and the vectors are of order n, as shown in Section C.4. Additionally, it is assumed that the interaction coefficients α_{ij} and α_{ji} among the isolated piles i, j are not affected by the presence of the other n-2 piles part of the pile group. As already mentioned, Eq. (B-10) directly applies for the case of horizontal or vertical loads. However, for rocking vibrations the displacements of the pile head must be expressed as a function of the rocking angle Ψ shown in Fig. B-2. Therefore, the unity terms in the vector I must be replaced by the proper distances (y_i and x_i) from the pile cap centroid to the pile head. For additional details refer to Section C.4 and El Naggar and Novak (1995a).

The interaction factors in Eqs. (B-3) and (B-10) can be calculated using analytical, numerical, and/or experimental methods (e.g., Kaynia and Kausel 1982; Dobry and Gazetas 1988; Makris and Gazetas 1992, 1993; Mylonakis 1995; Novak and El Sharnouby 1984; El et al. 1992). Here, for the sake of simplicity, the presentation is limited to the physically insightful interaction factors proposed by Dobry and Gazetas (1988) for two identical piles in a homogeneous half-space. The main assumptions of this approach are

- The response of the receiver pile to the oscillations of the source pile is assumed to be equal to the response of the free-field soil at the location of the receiver pile.
- The waves produced by the oscillating pile are assumed to emanate simultaneously from all perimeter points along the source pile length and wave scattering owing to other piles in the group is ignored (i.e., other piles are treated as invisible).
- As mentioned, piles in the wave trajectory are treated as invisible to the wave. Therefore, it is recommended to limit the minimum wavelength ($\lambda = 2\pi V_s/\Omega$) to approximately 6 times the pile diameter ($\lambda > 6d$). This imposes a limit on the maximum vibration frequency that should be analyzed with this method ($f_{max} < V_s/6d$ or $a_o \sim 1.0$).
- The number of piles in the group is small (e.g., less than 16 piles) and the soil is relatively soft, $E_p > 300E_s$, where E_p and E_s are the Young's modulus of the pile and soil, respectively.

Based on the given assumptions, the vertical α_{ij}^v and horizontal α_{ij}^h interaction factors are defined as follows:

$$\alpha_{ij}^{v} = \sqrt{\frac{R}{S}} e^{-\beta\Omega S/V_s} e^{-i\Omega S/V_s} \tag{B-11a}$$

$$\alpha_{ij}^{h} \approx \alpha_{ij}^{h}(0^\circ)\cos^2\theta + \alpha_{ij}^{h}(90^\circ)\sin^2\theta$$

where

$$\begin{aligned} \alpha_{ij}^{h}(0^\circ) &= \sqrt{\frac{R}{S}} e^{-\beta\Omega S/V_{La}} e^{-i\Omega S/V_{La}} \\ \alpha_{ij}^{h}(90^\circ) &= \sqrt{\frac{R}{S}} e^{-\beta\Omega S/V_s} e^{-i\Omega S/V_s} \end{aligned} \tag{B-11b}$$

$$V_{La} = \frac{3.4V_s}{\pi(1-\nu)} \tag{B-11c}$$

where α_{ij}^{v} and α_{ij}^{h} are vertical and horizontal interaction factors, respectively, S/R is the pile spacing to radius ratio between piles i and j, β is the soil material damping, θ is the angle between the direction of load action and the plane in which piles lie, and V_{La} is the so-called Lysmer's analog velocity.

In any case, independently of the approach used for calculating the pile interaction factors, the simplified method for calculating the dynamic impedance of a pile group discussed here is the same, and can be summarized as follows:

1. Calculate the impedance functions of each single pile (e.g., Novak 1974; Makris and Gazetas 1993), as discussed in Section B.1.
2. Calculate the pile to pile interaction factors (e.g., Dobry and Gazetas 1988; Mylonakis 1995; El et al. 1992; El Naggar and Novak 1995a).
3. Calculate the group impedance functions per Eq. (B-10). For additional details, refer to Dobry and Gazetas (1988) and El Naggar and Novak (1995b).
4. Repeat steps 1 to 3 for each vibration frequency of the machine, including startup and shutdown frequencies.

The application of these steps is illustrated in Appendix C; additional details are provided in the references included for each step.

REFERENCES

Dobry, R., and Gazetas, G. (1988). "Simple method for dynamic stiffness and damping of floating pile groups." *Géotechnique* 38(4), 557–574.

El, M. H., Kaynia, A. M., and Novák, M. (1992). *Interaction factors and the superposition method for pile group dynamic analysis*. Geotechnical Research Centre, University of Western Ontario, London, Ontario.

El Naggar, M. H., and Novak, M. (1995a). "Nonlinear lateral interaction in pile dynamics." *Soil Dyn. Earthquake Eng.* 14(2), 141–157.

El Naggar, M. H., and Novak, M. (1995b). "Effect of foundation nonlinearity on modal properties of offshore towers." *J. Geotech. Engrg. Div.* 121(9), 660–668.

Gazetas, G., and Makris, N. (1991). "Dynamic pile-soil-pile interaction, Part I: Analysis of axial vibration." *Earthquake Eng. Struct. Dyn.* 20, 115–132.

Kaynia, A. M., and Kausel, E. (1982). "Dynamic behavior of pile groups." 2nd International Conference on Numerical Methods in Offshore Piling, Austin, Texas, 509–532.

Lysmer, J. (1981). *SASSI: A system for analysis of soil-structure interaction.* Berkeley: University of California, Dept. of Civil Engineering.

Makris, N., and Gazetas, G. (1992). "Dynamic pile-soil-pile interaction, Part II: Lateral and seismic response." *Earthquake Eng. Struct. Dyn.* 21, 145–162.

Makris, N., and Gazetas, G. (1993). "Displacement phase differences in a harmonically oscillating pile." *Géotechnique* 43(1), 135–150.

Mylonakis, G. (1995). "Contributions to static and seismic analysis of piles and pile-supported bridge piers." http://catalog.hathitrust.org/api/volumes/oclc/40358980.html.

NIST (National Institute of Standards and Technology). (2012). *Soil-structure interaction for building structures.* GCR 12-917-21 U.S. Department of Commerce, Washington, DC.

Novak, M. (1974). "Dynamic stiffness and damping of piles." *Can. Geotech. J.* 11, 574–598.

Novak, M., and El Sharnouby, B. (1983). "Stiffness constants of single piles." *J. Geotech. Engrg. Div.* 109(7), 961–974.

Novak, M., and Sharnouby, B. E. (1984). "Evaluation of dynamic experiments on pile group." *J. Geotech. Engrg. Div.* 110(6), 738–756.

Poulos, H. G. (1971). "Behaviour of laterally loaded piles: II: Pile groups." *J. Soil Mech. and Found. Div.* 97 (SM5), 733–751.

APPENDIX C
DYNAMIC IMPEDANCE CALCULATION EXAMPLES

One of the objectives of this report is to present the methods and parameters required for calculating the response of machine foundations under dynamic loads. Therefore, this appendix begins (e.g., Sections C.1 and C.2) by showing the use of the dynamic soil impedances for calculating the subject response. A rigid foundation is used to keep the analysis simple and illustrate the fundamentals of the dynamic impedance method for modeling the soil response under dynamic loads (specifically harmonic).

The second part of this appendix is dedicated to illustrating the calculation of dynamic impedances using the procedures presented in Appendixes A and B. As already mentioned, the procedures covered in the subject appendixes are valid for modeling the response of rigid foundations. For instance, for block foundations and/or tabletop foundations, the foundation mat can be considered rigid, with the mass and mass moment of inertia lumped at the foundation center of gravity (CG). For additional application examples, the reader is referred to Novak and El Sharnouby (1983), Arya et al. (1984), Dobry and Gazetas (1988), Gazetas (1991), and Bowles (2007).

C.1 VERTICAL RESPONSE OF A RIGID FOUNDATION

This example illustrates the steps required for calculating the dynamic response of the foundation discussed in Section 5.1.1. For this purpose, consider the following soil parameters, foundation properties and machine load:

Description	Value
Operating frequency of machine	$f_o = 30\ Hz$ (i.e., $\Omega_0 = 60\pi\ rad/s$)
Mass of foundation	$m_f = 3770\ kg\ (4409\ lb)$
Mass of machine	$m_m = 2000\ kg\ (4409\ lb)$
Mass of rotor	$m_r = 500$ kg (1102 lb)
Rotor eccentricity	$e_r = 1.727 \times 10^{-4}\ m\ (5.67 \times 10^{-4}\ ft)$
Mass of machine foundation system	$m = 5770\ kg\ (12721\ lb)$
Operational unbalance load from rotor eccentricity	$F_0 = 3.068\ kN\ (0.69\ kip)$
Vertical impedance at operating frequency	$K_z(\Omega_o) = (1.591 + 3.715i) \times 10^5\ kN/m$

The motion equation and its corresponding solution are respectively given by Eqs. (5-2) and (5-3). After replacing the previous values in Eq. (5-3), the vertical displacement of the foundation is 8.196×10^{-6} m as shown in the following:

$$u_z(\Omega_0) = \frac{F_0}{K_z(\Omega) - \Omega^2 \mathrm{m}} = \frac{3068N}{(1.595 + 3.715i) \times 10^8 \dfrac{N}{\mathrm{m}} - (60\pi\ \mathrm{rad/s})^2 (5770\ \mathrm{kg})}$$

$$u_z(\Omega_0) = (-1.005 - 8.134i) \times 10^{-6} \mathrm{m}$$

$$|u_z(\Omega_0)| = \sqrt{1.005^2 + 8.134^2} \times 10^{-6} \mathrm{m} = 8.196 \times 10^{-6} \mathrm{m}$$

In these equations, it is important to note that the soil impedance is calculated at the operating frequency of the machine (i.e., $\Omega_o = 60\pi$ rad/s). The calculations must be repeated for each vibration frequency of the machine, including startup and shutdown frequencies. Additionally, for this simple foundation, the horizontal eccentricity between the CG of the machine foundation system and center of soil pressures or resistance (CR) is assumed to be zero, otherwise coupling between rocking and vertical vibration would have to be considered.

C.2 HORIZONTAL AND ROCKING RESPONSE OF A RIGID FOUNDATION

This example illustrates the steps required to calculate the dynamic response of the foundation discussed in Section 5.1.2. For this purpose, consider the following soil parameters, foundation properties and machine load:

Description	Value
Operating frequency of machine	$f_o = 30$ Hz (i.e., $\Omega_o = 60\pi$ rad/s)
Mass of machine foundation system	$m = 5770$ kg (12721 lb)
Mass moment of inertia of machine foundation system	$I_o = 1600000$ kg-m^2 (37970000 lb-ft^2)
Operational unbalance load from rotor eccentricity	$F_0 = 3.068$ kN (0.69 kip)
CR eccentricity with respect to CG	$z_c = 0.3$ m (1 ft)
Operational rocking moment from unbalance load	$M_0 = 0.767$ kN-m (0.52 kip-ft)
Horizontal impedance at operating frequency	$K_y(60\pi) = (1.091 + 1.857i) \times 10^5$ kN/m
Rocking impedance at operating frequency	$K_{rx}(60\pi) = (9.183 + 6.661i) \times 10^4$ kN-m/rad
Coupling impedance at operating frequency	$K_{yrx}(60\pi) = 0$

The equation of motion for the subject foundation is given by Eqs. (5-8) and (5-9), which can be written in matrix form (i.e., $[K]\{U\} = \{P\}$) as shown:

$$\begin{bmatrix} K_y - \Omega^2 m & K_{yrx} - K_y z_c \\ K_{yrx} - K_y z_c & (K_{rx} - 2K_{yrx} z_c + K_y z_c^2) - \Omega^2 I_{0X} \end{bmatrix} \begin{Bmatrix} u_y \\ \theta_x \end{Bmatrix} = \begin{Bmatrix} F_y \\ M_x \end{Bmatrix}$$

This matrix equation can be solved manually using Cramer's rule as illustrated next:

$$u_y = \frac{F_y \times K_{22} - M_x \times K_{12}}{K_{11} \times K_{22} - K_{12} \times K_{21}} = (-0.6736 - 1.304i) \times 10^{-5}\ \text{m}$$

$$\theta_x = \frac{M_x \times K_{11} - F_y \times K_{12}}{K_{11} \times K_{22} - K_{12} \times K_{21}} = (-2.245 + 1.41i) \times 10^{-8}\ \text{rad}$$

where

$$K_{11} = K_y - \Omega^2 m = (-9.591 \times 10^7 + 1.858i \times 10^8)\ \text{kg/s}^2$$
$$K_{12} = K_{21} = K_{yrx} - K_y z_c = (-3.273 - 5.573i) \times 10^7\ \text{N}$$
$$K_{22} = (K_{rx} - 2K_{yrx} z_c + K_y z_c^2) - \Omega^2 I_{0X} = (-5.675 \times 10^{10} + 8.333i \times 10^7)\ \text{N.m}$$

Therefore, the horizontal displacement and rotation of the machine foundation system CG are $|u_y| = 1.467 \times 10^{-5}$ m and $|\theta_x| = 2.651 \times 10^{-8}$ rad, which are calculated after replacing the numerical values for K_{11}, K_{12}, K_{22}, F_y, and M_x on the equations for u_y and θ_x. Finally, it should be noted that these calculations must be repeated for each vibration frequency of the machine, including startup and shutdown frequencies.

C.3 DYNAMIC IMPEDANCES OF SURFACE FOUNDATION

This example illustrates the steps required for calculating the dynamic impedances of foundations using the tables and figures provided in Appendix A. For this purpose, consider the following soil parameters, foundation properties, and machine operating frequency:

Soil Properties

Soil unit weight	$\gamma_s := 110\, pcf$	$\gamma_s = 17.3 \cdot \frac{kN}{m^3}$
Soil density	$\rho_s := \frac{\gamma_s}{g} = 3.4 \cdot \frac{lbf \cdot s^2}{ft^4}$	$\rho_s = 1762.0 \cdot \frac{kg}{m^3}$
Poisson's ratio	$v := 0.35$	
Shear wave velocity	$V_s := 600 \frac{ft}{s}$	$V_s = 182.9 \frac{m}{s}$
Lysmer's analog wave velocity	$V_{La} := \frac{3.4}{\pi \cdot (1 - v)} \cdot V_s = 999.0 \cdot \frac{ft}{s}$	$V_{La} = 304.5 \frac{m}{s}$
Dynamic shear modulus	$G := \frac{\gamma_s}{g} \cdot V_s^2 = 1230.8 \cdot ksf$	$G = 58.9 \cdot MPa$

Geometric Properties

Length of foundation base parallel to x-axis	$L_e := 16\, ft$	$L_e = 4.9\, m$
Width of foundation base parallel to y-axis	$B_e := 10\, ft$	$B_e = 3.0\, m$
Circumscribed rectangle length $2L$ parallel to x-axis	$L := \frac{L_e}{2} = 8.0 \cdot ft$	$L = 2.4\, m$
Circumscribed rectangle width $2B$ parallel to y-axis	$B := \frac{B_e}{2} = 5.0 \cdot ft$	$B = 1.5\, m$
Area of the actual foundation-soil contact surface	$A_b := 4 \cdot L \cdot B = 160.0 \cdot ft^2$	$A_b = 14.9 \cdot m^2$

Area moment of inertia about longitudinal (x) axis of the actual foundation-soil contact surface $I_{bx} := \frac{4}{3} L \cdot B^3 = 1333.3 \cdot ft^4$ $I_{bx} = 11.5 \cdot m^4$

Area moment of inertia about lateral (y) axis of the actual foundation-soil contact surface $I_{by} := \frac{4}{3} B \cdot L^3 = 3413.3 \cdot ft^4$ $I_{by} = 29.5 \cdot m^4$

Polar moment of inertia about (z) axis of the actual foundation-soil contact surface $J_b := I_{bx} + I_{by} = 4746.7 \cdot ft^4$ $J_b = 41.0 \cdot m^4$

Length to width aspect ratio $\frac{L}{B} = 1.6$

Actual area A_b to square foundation area ratio $X := \frac{A_b}{4 \cdot L^2} = 0.6$

Dynamic Stiffness and Dashpot Coefficients

Operating frequency of the machine $f_o := 30\ \text{Hz}$

Angular frequency of the machine $\Omega_o := 2\pi \cdot f_o = 188.5 \frac{\text{rad}}{\text{s}}$

Nondimensional frequency $a_o := \frac{B \cdot \Omega_o}{V_s} = 1.6$

Vertical Vibration Mode (z-axis)

Static stiffness (Table A-2) $K_z := \frac{2 \cdot G \cdot L}{(1-v)} \cdot (0.73 + 1.54 \cdot X^{0.75})$ $K_z = 801396.2 \cdot \frac{\text{kN}}{\text{m}}$

$= 54913.1 \cdot \frac{\text{kip}}{\text{ft}}$

Dynamic stiffness coefficient (Fig. A-2) $k_z := 0.75$

Damping coefficient (Fig. A-2) $c_z := 0.95$

Dynamic stiffness (Table A-2) $\bar{K}_z := k_z \cdot K_z = 41184.8 \cdot \frac{\text{kip}}{\text{ft}}$ $\bar{K}_z = 601047.1 \cdot \frac{\text{kN}}{\text{m}}$

Radiation dashpot coefficient (Table A-2) $C_z := \rho_s \cdot V_{La} \cdot A_b \cdot c_z$ $C_z = 7576.5 \cdot \text{kN} \cdot \frac{\text{s}}{\text{m}}$

$= 519.2 \cdot \text{kip} \cdot \frac{\text{s}}{\text{ft}}$

Lateral Vibration Mode (y-axis)

Static stiffness (Table A-2) $K_y := \frac{2 \cdot G \cdot L}{2 - v} \cdot (2 + 2.5 \cdot X^{0.85}) = 43880.9 \cdot \frac{\text{kip}}{\text{ft}}$ $K_y = 640394.0 \cdot \frac{\text{kN}}{\text{m}}$

Dynamic stiffness coefficient (Fig. A-2) $k_y := 0.95$

Damping coefficient (Fig. A-2) $c_y := 0.95$

Dynamic stiffness (Table A-2) $\bar{K}_y := k_y \cdot K_y = 41686.9 \cdot \frac{\text{kip}}{\text{ft}}$ $K_y = 608374.3 \cdot \frac{\text{kN}}{\text{m}}$

Radiation dashpot coefficient (Table A-2) $C_y := \rho_s \cdot V_s \cdot A_b \cdot c_y = 311.8 \cdot \text{kip} \cdot \frac{\text{s}}{\text{ft}}$ $C_y = 4550.4 \cdot kN \cdot \frac{\text{s}}{\text{m}}$

Longitudinal Vibration Mode (x-axis)

Static stiffness (Table A-2) $K_x := K_y - \frac{0.2}{0.75 - v} \cdot G \cdot L \cdot \left(1 - \frac{B}{L}\right) = 42034.7 \cdot \frac{k}{J}$ $K_x = 613450.6 \cdot \frac{\text{kN}}{\text{m}}$

Dynamic stiffness coefficient (Table A-2) $k_x := 1.0$

Dynamic stiffness (Table A-2) $\bar{K}_x := k_x \cdot K_x = 42034.7 \cdot \frac{\text{kip}}{\text{ft}}$ $\bar{K}_x = 613450.6 \cdot \frac{\text{kN}}{\text{m}}$

Radiation dashpot coefficient (Table A-2) $C_x := \rho_s \cdot V_s \cdot A_b = 328.2 \cdot \text{kip} \cdot \frac{\text{s}}{\text{ft}}$ $C_x = 4789.9 \cdot \text{kN} \cdot \frac{\text{s}}{\text{m}}$

Rocking, rx (around longitudinal x-axis)

Static stiffness (Table A-2) $K_{rx} := \frac{G}{1 - v} \cdot I_{bx}{}^{0.75} \cdot \left(\frac{L}{B}\right)^{0.25} \cdot \left(2.4 + 0.5 \cdot \frac{B}{L}\right) = 1274617.6 \cdot \frac{kip \cdot ft}{rad}$ $K_{rx} = 1728149.4 \cdot \frac{\text{kN} \cdot \text{m}}{\text{rad}}$

Dynamic stiffness coefficient (Table A-2) $k_{rx} := (1 - 0.2 \cdot a_o) = 0.7$

Damping coefficient (Fig. A-2) $c_{rx} := 0.45$

Dynamic stiffness (Table A-2) $\bar{K}_{rx} := k_{rx} \cdot K_{rx} = 874184.7 \cdot \frac{\text{kip} \cdot \text{ft}}{\text{rad}}$ $\bar{K}_{rx} = 1185235.3 \cdot \frac{\text{kN} \cdot \text{m}}{\text{rad}}$

Radiation dashpot coefficient (Table A-2)
$$C_{rx} := \rho_s \cdot V_{La} \cdot I_{bx} \cdot c_{rx} = 2049.3 \cdot \frac{\text{kip} \cdot \text{s} \cdot \text{ft}}{\text{rad}} \qquad C_{rx} = 2778.5 \cdot \frac{\text{kN} \cdot \text{s} \cdot \text{m}}{\text{rad}}$$

Rocking, ry (around lateral *y*-axis)

Static stiffness (Table A-2)
$$K_{ry} := \frac{G}{1-\upsilon} \cdot I_{by}{}^{0.75} \cdot \left[3 \cdot \left(\frac{L}{B}\right)^{0.15}\right] = 2722070.7 \cdot \frac{\text{kip} \cdot \text{ft}}{\text{rad}} \qquad K_{ry} = 3690632.3 \cdot \frac{\text{kN} \cdot \text{m}}{\text{rad}}$$

Dynamic stiffness coefficient (Table A-2)
$$k_{ry} := (1 - 0.3 \cdot a_o) = 0.5$$

Damping coefficient (Fig. A-2)
$$c_{ry} := 0.55$$

Dynamic stiffness (Table A-2)
$$\bar{K}_{ry} := k_{ry} \cdot K_{ry} = 1439325.1 \cdot \frac{\text{kip} \cdot \text{ft}}{\text{rad}} \qquad \bar{K}_{ry} = 1951462.8 \cdot \frac{\text{kN} \cdot \text{m}}{\text{rad}}$$

Radiation dashpot coefficient (Table A-2)
$$C_{ry} := \rho_s \cdot V_{La} \cdot I_{by} \cdot c_{ry} = 6412.0 \cdot \frac{\text{kip} \cdot \text{s} \cdot \text{ft}}{\text{rad}} \qquad C_{ry} = 8693.5 \cdot \frac{\text{kN} \cdot \text{s} \cdot \text{m}}{\text{rad}}$$

Torsional

Static stiffness (Table A-2)
$$K_t := G \cdot J_b{}^{0.75} \cdot \left[4 + 11 \cdot \left(1 - \frac{B}{L}\right)^{10}\right] = 2815831.7 \cdot \frac{\text{kip} \cdot \text{ft}}{\text{rad}} \qquad K_t = 3817755.1 \cdot \frac{\text{kN} \cdot \text{m}}{\text{rad}}$$

Dynamic stiffness coefficient (Table A-2)
$$k_t := (1 - 0.14 \cdot a_o) = 0.8$$

Damping coefficient (Fig. A-2)
$$c_t := 0.65$$

Dynamic stiffness (Table A-2)
$$\bar{K}_t := k_t \cdot K_t = 2196597.9 \cdot \frac{\text{kip} \cdot \text{ft}}{\text{rad}} \qquad \bar{K}_t = 2978186.9 \cdot \frac{\text{kN} \cdot \text{m}}{\text{rad}}$$

Radiation dashpot coefficient (Table A-2)
$$C_t := \rho_s \cdot V_s \cdot J_b \cdot c_t = 6329.1 \cdot \frac{\text{kip} \cdot \text{s} \cdot \text{ft}}{\text{rad}} \qquad C_t = 8581.1 \cdot \frac{\text{kN} \cdot \text{s} \cdot \text{m}}{\text{rad}}$$

It must be noted that these calculations must be repeated for each vibration frequency of the machine, including startup and shutdown frequencies.

C.4 PILE GROUP EFFECTS

The purpose of this example is to illustrate the calculation of the dynamic impedance of a 3 by 3 pile group, as shown in Fig. C-1. The dynamic interaction method presented in Section B.2.2 is used for this purpose. In particular, axial and rocking impedances are calculated; for additional examples, the reader is referred to Dobry and Gazetas (1988). The following pile, soil, and machine data are used to illustrate the various steps.

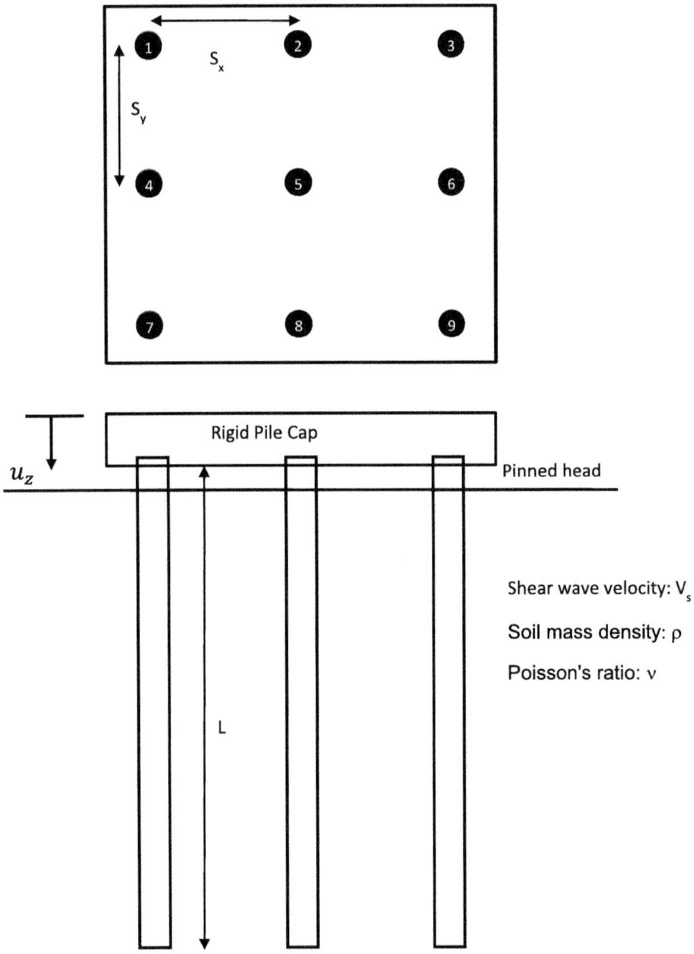

Fig. C-1. Pile group geometry

Pile Group Properties

Single pile radius	$R := 1\ \text{ft}$	$R = 0.3\ \text{m}$
Pile length	$L := 30 \cdot R := 30 \cdot \text{ft}$	$L = 9.14\ \text{m}$
Young's modulus of the pile	$E_p := 4400\ \text{ksi}$	$E_p = 30.34 \cdot \text{GPa}$
Number of piles in the longitudinal and lateral directions	$n_x := 3$	$n_y = 3$
Pile spacing in the longitudinal direction	$S_x := 10 \cdot R := 10 \cdot \text{ft}$	$S_x = 3.05\ \text{m}$
Pile spacing in the lateral direction	$S_y := 8 \cdot R := 8 \cdot \text{ft}$	$S_y = 2.44\ \text{m}$
Shear wave velocity	$V_s := 400 \frac{\text{ft}}{\text{s}}$	$V_s = 121.92 \frac{\text{m}}{\text{s}}$
Soil density	$\rho := \frac{120}{g} \text{pcf} = 3.73 \cdot \frac{\text{lbf} \cdot \text{s}^2}{\text{ft}^4}$	$\rho = 1.92 \times 10^3 \cdot \frac{\text{kg}}{\text{m}^3}$
Poisson's ratio	$\text{v} := 0.3$	$V_{La} = 188.5 \frac{\text{m}}{\text{s}}$
Lysmer's analog wave velocity	$V_{La} := \frac{3.4 \cdot V_s}{\pi \cdot (1\text{-v})} = 618.43 \cdot \frac{\text{ft}}{\text{s}}$	
Shear modulus of the soil	$G_s := V_s^2 \cdot \rho = 596.75 \cdot \text{ksf}$	$G_S := 28.57 \cdot \text{MPa}$
Young's modulus of the soil	$E_S := G_S \cdot 2(1 + \text{v}) = 1552 \cdot \text{ksf}$	$E_S := 74.29 \cdot \text{MPa}$
Material damping of the soil	$\beta := 0.05$	$\frac{E_p}{E_s} = 408.36$
Depth of homogeneous stratum	$H := 100\ \text{ft}$	$H := 30.48\ \text{m}$
Operating frequency of the machine	$f_o := 30\ \text{Hz}$	
Circular frequency of the machine	$\Omega_o := 2\pi \cdot f_o = 188.5 \cdot \frac{\text{rad}}{\text{s}}$	
Frequency limit (see Section B.2.2)	$f_{max} := \frac{V_s}{12 \cdot R} = 33.33 \frac{1}{\text{s}}$	
	$f_o < f_{max}$, OK	

Step 1: Calculate the nondimensional frequency a_0 from the machine excitation and the fundamental frequencies of the soil layer.

Nondimensional frequency $\quad a_o := \dfrac{2 \cdot R \cdot \Omega_o}{V_S} = 0.94 \quad$ $a_0 < 1$, therefore OK

Circular frequency of soil stratum in the horizontal direction $\quad \omega_{su} := \dfrac{\pi \cdot V_s}{2 \cdot H} = 6.3 \cdot \dfrac{\text{rad}}{\text{s}} \quad < \Omega_o = 188.5 \cdot \dfrac{\text{rad}}{\text{s}}$

Circular frequency of soil stratum in the vertical direction $\quad \omega_{sv} := \dfrac{\pi \cdot V_s}{2 \cdot H} \cdot \sqrt{\dfrac{2(1-\nu)}{1-2 \cdot \nu}} \quad < \Omega_o = 188.5 \cdot \dfrac{\text{rad}}{\text{s}}$

$$= 11.8 \cdot \frac{\text{rad}}{\text{s}}$$

As can be seen the excitation frequency is greater than the frequencies of the soil layer; therefore, damping corrections are not required.

Step 2: Use Eq. (B-1), Table B-2, and Fig. B-1(b) to calculate the vertical and rocking impedances of a single pile. Note that the piles are pinned, thus $K_{rx} = K_{ry} = 0$.

Dynamic impedance for a single pile, Eq. (B-1) $\quad K_z(a_o) = k_z(a_o) + i\Omega c_z(a_o)$

Pile area $\quad A_p := \pi \cdot R^2 = 3.14 \cdot \text{ft}^2 \quad A_p = 0.29\ \text{m}^2$

Pile slenderness and stiffness ratios $\quad \lambda := \dfrac{L}{R} = 30 \quad \dfrac{E_p}{G_s} = 1061.7$

Dimensionless stiffness and damping functions, Fig. B-2(b) $\quad f_{v1} := 0.02 \quad f_{v2} := 0.055$

Vertical stiffness coefficient, Table B-2 $\quad k_z := \dfrac{E_p \cdot A_p}{R} \cdot f_{v1} \quad k_z = 5.81 \times 10^5 \cdot \dfrac{\text{kN}}{\text{m}}$

$$= 3.98 \times 10^4 \cdot \frac{\text{kip}}{\text{ft}}$$

Vertical damping coefficient, Table B-2 $\quad c_z := \dfrac{E_p \cdot A_p}{V_s} \cdot f_{v2} \quad c_z = 3.99 \times 10^3 \cdot \dfrac{\text{kN}}{\text{m}} \cdot \text{s}$

$$= 273.7 \cdot \frac{\text{kip}}{\text{ft}} \cdot \text{s}$$

Dynamic pile impedance for vertical vibration $\quad K_z(\Omega) := k_z + i \cdot \Omega \cdot c_z$

$$K_z(\Omega_o) = (39810 + 51590i) \cdot \frac{\text{kip}}{\text{ft}} \qquad K_Z(\Omega_o) = (580987 + 752905i) \cdot \frac{\text{kN}}{\text{m}}$$

Dynamic pile impedance for rocking vibration (pinned head condition) $K_{rx} := 0$ $K_{ry} := 0$

Step 3: Use Eq. [B-11(a)] to calculate the interaction factors among the different piles of the group. For illustration purposes, the following interaction factors are calculated; similar calculations are required for all the piles:

$$\alpha_v = \sqrt{\frac{R}{S}} \cdot e^{\frac{-\beta \cdot \Omega \cdot S}{V_S}} \cdot e^{\frac{-i \cdot \Omega \cdot S}{V_s}}$$

where

$$|\alpha_v| = \sqrt{\frac{R}{S}} \cdot e^{\frac{-\beta \cdot \Omega \cdot S}{V_S}}$$

or

$$\alpha_v = |\alpha_v| (cos(\phi) - i \cdot sin(\phi))$$

where

$$\phi = \frac{\Omega S}{V_S}$$

Displacement of pile 1 head from a unit load applied on pile 2 (α_{12})

$S := S_x = 10 \cdot \text{ft}$ $S = 3.05\ \text{m}$

$$\alpha_{12} := \left(\frac{S}{R}\right)^{-0.5} \cdot e^{(-\beta \cdot \Omega_o \cdot S) \div V_s} \cdot e^{(-i \cdot \Omega_o \cdot S) \div V_s} = 0.2498i$$

Displacement of pile 1 head from a unit load applied on pile 3 (α_{13})

$S := 2S_x = 20 \cdot \text{ft}$ $S = 6.1\ \text{m}$

$$\alpha_{13} := \sqrt{\frac{R}{S}} \cdot e^{(-\beta \cdot \Omega_o \cdot S) \div V_S} \cdot e^{(-i \cdot \Omega_o \cdot S) \div V_S} = -0.1396$$

Displacement of pile 1 head from a unit load applied on pile 4 (α_{14})

$S := S_y = 8 \cdot \text{ft}$ $S = 2.44\ \text{m}$

$$\alpha_{14} := \left(\frac{S}{R}\right)^{-0.5} \cdot e^{(-\beta \cdot \Omega_o \cdot S) \div V_S} \cdot e^{(-i \cdot \Omega_o \cdot S) \div V_S} = -0.2369 + 0.1721i$$

Displacement of pile 1 head from a unit load applied on pile 5 (α_{15})

$S := \sqrt{(S_x)^2 + (S_y)^2} = 12.81 \cdot \text{ft}$ $S = 3.9\ \text{m}$

$$\alpha_{15} := \left(\frac{S}{R}\right)^{-0.5} \cdot e^{(-\beta\cdot\Omega_o\cdot S)\div V_S} \cdot e^{(-i\cdot\Omega_o\cdot S)\div V_S} = 0.2003 + 0.0508i$$

Displacement of pile 1 head from a unit load applied on pile 9 (α_{19})

$$S := \sqrt{(2S_x)^2 + (2S_y)^2} = 25.61\cdot\text{ft} \qquad S = 7.81\ \text{m}$$

$$\alpha_{19} := \left(\frac{S}{R}\right)^{-0.5} \cdot e^{(-\beta\cdot\Omega_o\cdot S)\div V_S} \cdot e^{(-i\cdot\Omega_o\cdot S)\div V_S} = 0.095 + 0.0515i$$

For illustration purposes, Fig. C-2 shows the variation of the vertical interaction factors of pile i from the vibration of pile j (i.e., α_{vij}). As can be seen, the dynamic interaction factors of pile i from pile j depend strongly on the vibration frequency. In particular, the vibration of pile j can increase, reduce, or not affect the displacements of pile i, depending on the frequency and distance S between the piles (i.e., positive or negative interference). Therefore, dynamic group effect factors can be less or greater than one, depending on the frequency, whereas static group effect factors are always less than 1. Additionally, the influence of pile j on pile i reduces (it is attenuated) as the spacing and frequency increase. In fact, the subject influence becomes negligible for $S/d > 30$.

Step 4: Assemble the dynamic interactions factors to form the matrix (A), as given by Eq. (B-3), and calculate its corresponding inverse.

Real part of the dynamic interaction matrix:

$$Re(A) = \begin{pmatrix} 1 & 0 & -0.1396 & -0.2369 & 0.2003 & -0.097 & 0.053 & -0.1271 & 0.095 \\ 0 & 1 & 0 & 0.2003 & -0.2369 & 0.2003 & -0.1271 & 0.053 & -0.1271 \\ -0.1396 & 0 & 1 & -0.097 & 0.2003 & -0.2369 & 0.095 & -0.1271 & 0.053 \\ -0.2369 & 0.2003 & -0.097 & 1 & 0 & -0.1396 & -0.2369 & 0.2003 & -0.097 \\ 0.2003 & -0.2369 & 0.2003 & 0 & 1 & 0 & 0.2003 & -0.2369 & 0.2003 \\ -0.097 & 0.2003 & -0.2369 & -0.1396 & 0 & 1 & -0.097 & 0.2003 & -0.2369 \\ 0.053 & -0.1271 & 0.095 & -0.2369 & 0.2003 & -0.097 & 1 & 0 & -0.1396 \\ -0.1271 & 0.053 & -0.1271 & 0.2003 & -0.2369 & 0.2003 & 0 & 1 & 0 \\ 0.095 & -0.1271 & 0.053 & -0.097 & 0.2003 & -0.2369 & -0.1396 & 0 & 1 \end{pmatrix}$$

Imaginary part of the dynamic interaction matrix:

$$Im(A) = \begin{pmatrix} 0 & 0.2498 & 0 & 0.1721 & 0.0508 & 0.0861 & -0.1631 & -0.0751 & 0.0515 \\ 0.2498 & 0 & 0.2498 & 0.0508 & 0.1721 & 0.0508 & -0.0751 & -0.1631 & -0.0751 \\ 0 & 0.2498 & 0 & 0.0861 & 0.0508 & 0.1721 & 0.0515 & -0.0751 & -0.1631 \\ 0.1721 & 0.0508 & 0.0861 & 0 & 0.2498 & 0 & 0.1721 & 0.0508 & 0.0861 \\ 0.0508 & 0.1721 & 0.0508 & 0.2498 & 0 & 0.2498 & 0.0508 & 0.1721 & 0.0508 \\ 0.0861 & 0.0508 & 0.1721 & 0 & 0.2498 & 0 & 0.0861 & 0.0508 & 0.1721 \\ -0.1631 & -0.0751 & 0.0515 & 0.1721 & 0.0508 & 0.0861 & 0 & 0.2498 & 0 \\ -0.0751 & -0.1631 & -0.0751 & 0.0508 & 0.1721 & 0.0508 & 0.2498 & 0 & 0.2498 \\ 0.0515 & -0.0751 & -0.1631 & 0.0861 & 0.0508 & 0.1721 & 0 & 0.2498 & 0 \end{pmatrix}$$

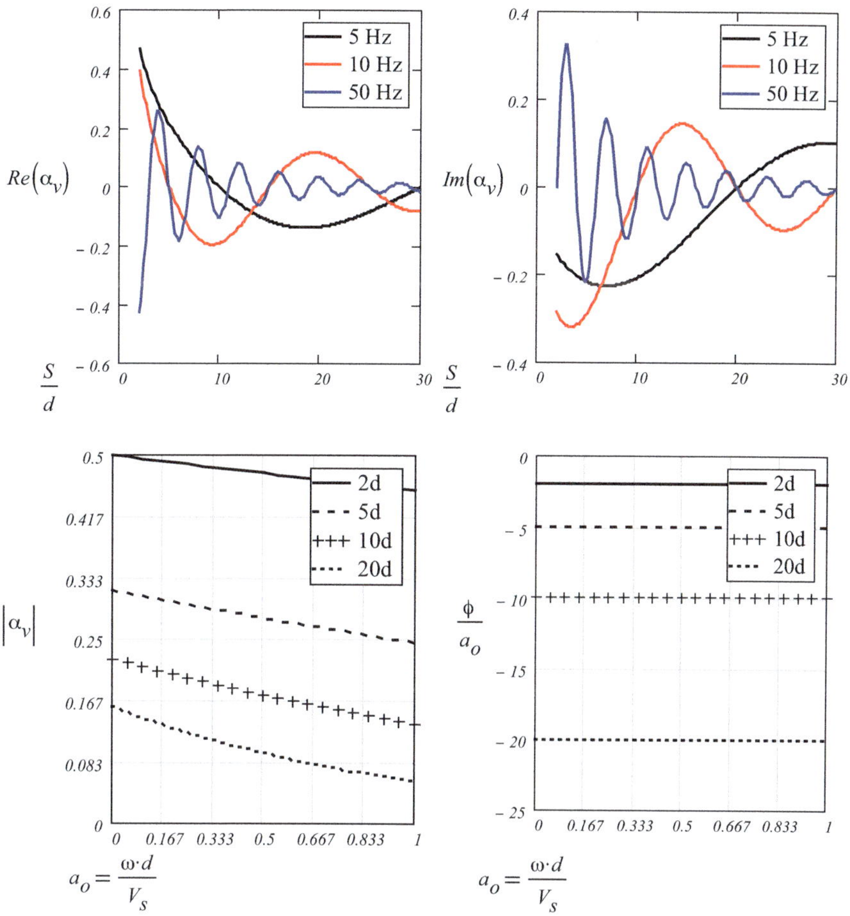

Fig. C-2. Vertical interaction factors as functions of pile spacing and excitation frequency

Real part of the inverse of the dynamic interaction matrix:

$$Re(A^{-1}) = \begin{pmatrix} 1.0637 & -0.1822 & 0.201 & 0.3069 & -0.3565 & 0.1847 & 0.0831 & -0.0045 & 0.0398 \\ -0.1822 & 0.9781 & -0.1822 & -0.3243 & 0.2347 & -0.3243 & -0.0045 & 0.1201 & -0.0045 \\ 0.201 & -0.1822 & 1.0637 & 0.1847 & -0.3565 & 0.3069 & 0.0398 & -0.0045 & 0.0831 \\ 0.3069 & -0.3243 & 0.1847 & 1.2081 & -0.3761 & 0.3083 & 0.3069 & -0.3243 & 0.1847 \\ -0.3565 & 0.2347 & -0.3565 & -0.3761 & 1.1828 & -0.3761 & -0.3565 & 0.2347 & -0.3565 \\ 0.1847 & -0.3243 & 0.3069 & 0.3083 & -0.3761 & 1.2081 & 0.1847 & -0.3243 & 0.3069 \\ 0.0831 & 0.0045 & 0.0398 & 0.3069 & 0.3565 & 0.1847 & 1.0637 & -0.1822 & 0.201 \\ -0.0045 & 0.1201 & -0.0045 & -0.3243 & 0.2347 & -0.3243 & -0.1822 & 0.9781 & -0.1822 \\ 0.0398 & -0.0045 & 0.0831 & 0.1847 & -0.3565 & 0.3069 & 0.201 & -0.1822 & 1.0637 \end{pmatrix}$$

Imaginary part of the inverse of the dynamic interaction matrix:

$$Im(A^{-1}) = \begin{pmatrix} 0.078 & -0.3383 & 0.0799 & 0.0108 & -0.2718 & 0.0398 & 0.0935 & 0.0386 & -0.0534 \\ -0.3383 & 0.2772 & -0.3383 & -0.3047 & 0.2255 & -0.3047 & 0.0386 & 0.2018 & 0.0386 \\ 0.0799 & -0.3383 & 0.078 & 0.0398 & -0.2718 & -0.0108 & -0.0534 & 0.0386 & 0.0935 \\ -0.0108 & -0.3047 & 0.0398 & 0.1376 & -0.4732 & 0.1685 & -0.0108 & -0.3047 & 0.0398 \\ -0.2718 & 0.2255 & -0.2718 & -0.4732 & 0.5042 & -0.4732 & -0.2718 & 0.2255 & -0.2718 \\ 0.0398 & -0.3047 & -0.0108 & 0.1685 & -0.4732 & 0.1376 & 0.0398 & -0.3047 & -0.0108 \\ 0.0935 & 0.0386 & -0.0534 & -0.0108 & -0.2718 & 0.0398 & 0.078 & -0.3383 & 0.0799 \\ 0.0386 & 0.2018 & 0.0386 & -0.3047 & 0.2255 & -0.3047 & -0.3383 & 0.2772 & -0.3383 \\ -0.0534 & 0.0386 & 0.0935 & 0.0398 & -0.2718 & -0.0108 & 0.0799 & -0.3383 & 0.078 \end{pmatrix}$$

Step 5: Calculate the terms in the vector I and assemble the matrix K, as given by Eq. (B-3).

Pile head displacement under vertical load (I_v):

$$I_v^T = (1\ 1\ 1\ 1\ 1\ 1\ 1\ 1\ 1)$$

y-coordinate of each pile from the CG of the group (I_{rx}):

$$I_{rx}^T = (-8\ -8\ -8\ 0\ 0\ 0\ 8\ 8\ 8)\cdot \text{ft}$$

$$I_{rx}^T = (-2.438\ -2.438\ -2.438\ 0\ 0\ 0\ 2.438\ 2.438\ 2.438)\cdot \text{ma}$$

x-coordinate of each pile from the CG of the group (I_{ry}):

$$I_{ry}^T = (-10\ 0\ 10\ -10\ 0\ 10\ -10\ 0\ 10)\cdot \text{ft}$$

$$I_{ry}^T = (-3.048\ 0\ 3.048\ -3.048\ 0\ 3.048\ -3.048\ 0\ 3.048)\cdot \text{m}$$

Pile impedance matrix (K)

$$K := K_z(\Omega_o)\cdot identity(n_x \cdot n_y)$$

$$K_z(\Omega_o) = (3.98\times 10^4 + 5.16i\times 10^4)\cdot \frac{\text{kip}}{\text{ft}}$$

$$K_Z(\Omega_o) = (5.81\times 10^5 + 7.53i\times 10^5)\cdot \frac{\text{kN}}{\text{m}}$$

Step 6: Calculate the group impedances per Eq. (B-10).

Vertical impedance of the pile group for $\Omega_o = 188.5\frac{1}{\text{s}}$

$$K_{GV} := I_v^T \cdot A^{-1}\cdot K \cdot I_v = (5.87\times 10^5 + 2.38i\times 10^5)\cdot \frac{\text{kip}}{\text{ft}}$$

$$K_{GV} := (8.56\times 10^6 + 3.47i\times 10^6)\cdot \frac{\text{kN}}{\text{m}}$$

Rocking rx impedance of the pile group for $\Omega_o = 188.5\frac{1}{\text{s}}$

$$K_{Grx} := n_x \cdot n_y \cdot K_{rx} + I_{rx}^T \cdot A^{-1} \cdot K \cdot I_{rx} = (2.03 \times 10^7 + 9.95i \times 10^6) \cdot \frac{\text{kip} \cdot \text{fit}}{\text{rad}}$$

$$K_{Grx} = (2.75 \times 10^7 + 1.35i \times 10^7) \cdot \frac{\text{kN} \cdot \text{m}}{\text{rad}}$$

Rocking ry impedance of the pile group for $\Omega_o = 188.5 \frac{1}{\text{s}}$

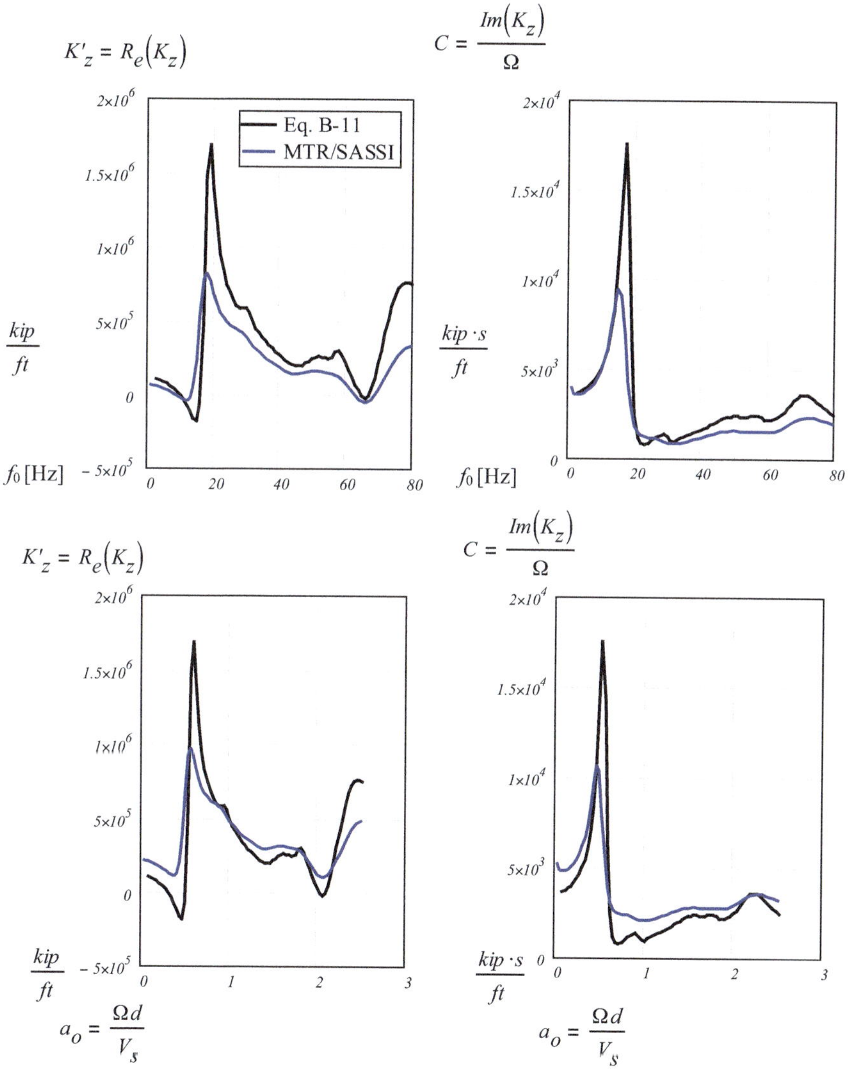

Fig. C-3. Vertical stiffness and damping coefficients for a 3 by 3 pile group

$$K_{Gry} := n_x \cdot n_y \cdot K_{rx} + I_{ry}^T \cdot A^{-1} \cdot K \cdot I_{ry} = (2.49 \times 10^7 + 3.35i \times 10^7) \cdot \frac{\text{kip} \cdot \text{fit}}{\text{rad}}$$

$$K_{Gry} = (3.38 \times 10^7 + 4.54i \times 10^7) \cdot \frac{\text{kN} \cdot \text{m}}{\text{rad}}$$

Step 7: Repeat steps 1 to 6 for each vibration frequency of the machine, including startup and shutdown frequencies.

Fig. C-3 shows the variation with frequency of the vertical stiffness and damping coefficients for the subject 3 by 3 pile group. Results calculated using MTR/SASSI (2015) are also included for comparison purposes. As can be seen, the two sets of results follow similar trends, despite the simplicity and limitations of interaction method, discussed in Section B.2.2. Nevertheless, it must be noted that for large pile groups, larger differences are expected (see C.5).

C.5 DYNAMIC IMPEDANCE OF A LARGE PILE GROUP

This example illustrates the calculation of the vertical and horizontal impedances for a 6 by 6 pile group embedded in a deep layered soil profile. The computer codes PILES (Kaynia 1982), MTR/SASSI (2015), and DYNA6 (El Naggar et al. 2011) are used for this purpose. The pile group properties are shown in Fig. C-4. As can be seen, the piles have uniform solid circular cross section of diameter $d = 0.456\ m$ and embedment length of $L = 15\ m$ $(L/d = 33)$. The pile spacing is $S = 1.2\ m$ center-to-center $(S/d = 2.6)$. The soil velocity profile is also shown in Fig. C-4 and detailed in Table C-1. The dynamic impedances are calculated in the frequency range from 6 Hz to 80 Hz. The range of nondimensional frequencies ($a_o = \omega d/V_s$) is approximately from 0.1 to 1.2, where the weighted average shear wave velocity along the pile length is used $(V_s = 194\ m/s)$. Thus, the nondimensional frequency is within the range deemed acceptable for the dynamic interaction factor method used by DYNA6; see Section B.2.2.

To model the dynamic behavior of a pile group in a layered soil profile, the computer code PILES uses the dynamic flexibility matrix of the soil along with the dynamic stiffness and flexibility matrices of the piles, as detailed by Kaynia (1982). The main assumptions of this approach are 1) the soil layers are modeled as viscoelastic materials, 2) the piles are modeled as linear elastic beams, and 3) the bond between the soil and piles is perfect (i.e., no slip), although frictional effects from torsion and bending are ignored. Overall, the Kaynia (1982) approach is considered highly accurate and is deemed herein as the benchmark for other solutions.

The computer code MTR/SASSI (2015) uses the finite element method and ring load solution in layered soil system to model the behavior of pile

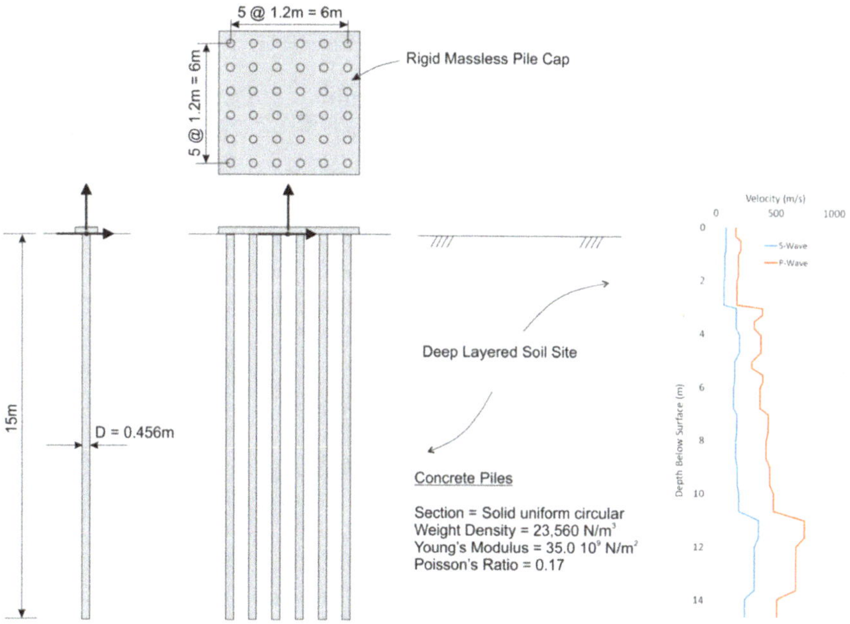

Fig. C-4. Single pile and pile group configuration, including geometric and material properties

groups. The methodology is based on derivation of the dynamic compliance matrix for the pile group using traction forces and displacements on individual pile shaft and tip calculated from axisymmetric ring load solution. The fully populated compliance matrix is then inverted to calculate the dynamic impedance matrix for the pile group. The impedance matrix is then assembled with the finite element model of the pile cap structure system and solved in the frequency domain for each vibration frequency. Because dynamic group factors are calculated rigorously using the finite element method, no interaction factors and/or superposition are required to obtain the pile group response. This approach assumes viscoelastic soil behavior, linear elastic piles, and full bond between the piles and soil; for additional details regarding the subject solution, the reader is referred to Tabatabaie (2015, 2017). MTR/SASSI (2015) finite-element model for this foundation is shown in Fig. C-5. As can be seen, the pile cap is modeled as rigid. Nevertheless, the approach presented here can handle piles with both rigid or flexible pile caps and can be used with any foundation/superstructure configuration.

The computer code DYNA6 uses the dynamic interaction factor method to calculate the dynamic impedance of pile groups in layered soil profiles,

Table C-1. Soil Profile Properties

Thickness, (m)	Shear Velocity, (m/s)	Mass Density, (kg/m^3)	Shear Modulus, (Pa)	Damping Ratio	Poisson's Ratio
0.25	84	1900	1.340640E+07	0.020	0.320
0.25	84	1900	1.340640E+07	0.020	0.320
0.25	84	1900	1.340640E+07	0.020	0.400
0.25	84	1900	1.340640E+07	0.020	0.400
0.25	75	1900	1.068750E+07	0.020	0.400
0.25	75	1900	1.068750E+07	0.020	0.400
0.25	75	1900	1.068750E+07	0.020	0.400
0.25	75	1900	1.068750E+07	0.020	0.400
0.25	71	1900	9.577900E+06	0.020	0.400
0.25	71	1900	9.577900E+06	0.020	0.400
0.25	71	1900	9.577900E+06	0.020	0.400
0.25	71	1900	9.577900E+06	0.020	0.400
0.50	170	1900	5.491000E+07	0.020	0.380
0.50	170	1900	5.491000E+07	0.020	0.300
1.00	200	1900	7.600000E+07	0.020	0.300
0.50	160	1900	4.864000E+07	0.020	0.300
0.50	160	1900	4.864000E+07	0.020	0.400
1.00	151	1900	4.332190E+07	0.020	0.400
1.00	179	1900	6.087790E+07	0.020	0.400
1.00	172	1900	5.620960E+07	0.020	0.400
1.00	186	1900	6.573240E+07	0.020	0.400
1.00	200	1900	7.600000E+07	0.020	0.400
1.00	365	1900	2.531275E+08	0.020	0.350
2.00	330	1900	2.069100E+08	0.020	0.350
1.00	250	1900	1.187500E+08	0.020	0.350
Half Space	250	1900	1.187500E+08	0.020	0.350

as described by El Naggar et al. (2011). In this approach, the impedance functions of single piles and the interaction factors are calculated first, then the group impedance functions are computed. The stiffness and damping constants of individual piles are calculated using expressions given by Novak (1974). The interaction factors are calculated using the dynamic interaction factors reported by Kaynia (1982). The impedance functions for a pile group of N piles are then given by the following equations proposed by El Naggar and Novak (1995):

$$K_v^G = \bar{k}_v \sum_{i=1}^{n} \sum_{j=1}^{n} \varepsilon_{ij}^v$$

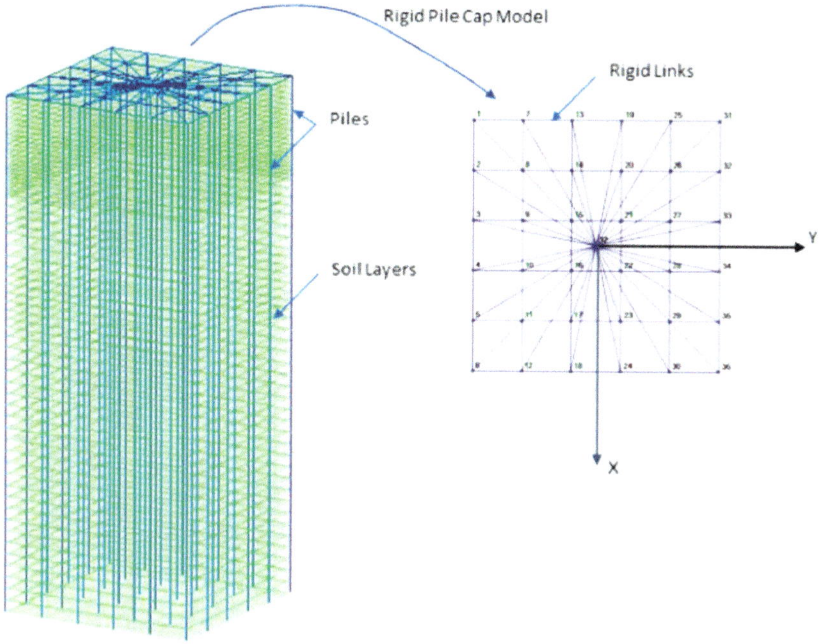

Fig. C-5. Finite element model of the pile group, including soil layers

$$K_h^G = \bar{k}_h \sum_{i=1}^{n} \sum_{j=1}^{n} \varepsilon_{2i-1,2j-1}^h$$

$$K_r^G = \bar{k}_h \sum_{i=1}^{n} \sum_{j=1}^{n} \varepsilon_{2i,2j}^h$$

$$K_c^G = \bar{k}_h \sum_{i=1}^{n} \sum_{j=1}^{n} \varepsilon_{2i-1,2j}^h$$

where K_v^G, K_h^G, K_r^G, and K_c^G are the vertical, horizontal, rocking, and coupling group stiffness, respectively. In Eq. (5-14), $\bar{k}_v$ is the static vertical stiffness of the single pile, $[\varepsilon^v]=[\alpha]_v^{-1}$, where α_{ij}^v = complex interaction factors between piles i and j, $\alpha_{ii}^v = \bar{k}_v / K_v$, and K_v is the complex vertical impedance function of the single pile. Similarly, $\bar{k}_h$ is the static horizontal stiffness of the pile $[\varepsilon^h]=[\alpha]_h^{-1}$ where α_{ij}^h = complex interaction coefficients for the horizontal translations and rotations. The formulation of the $[\alpha]_h$ can be found in El Naggar and Novak (1995).

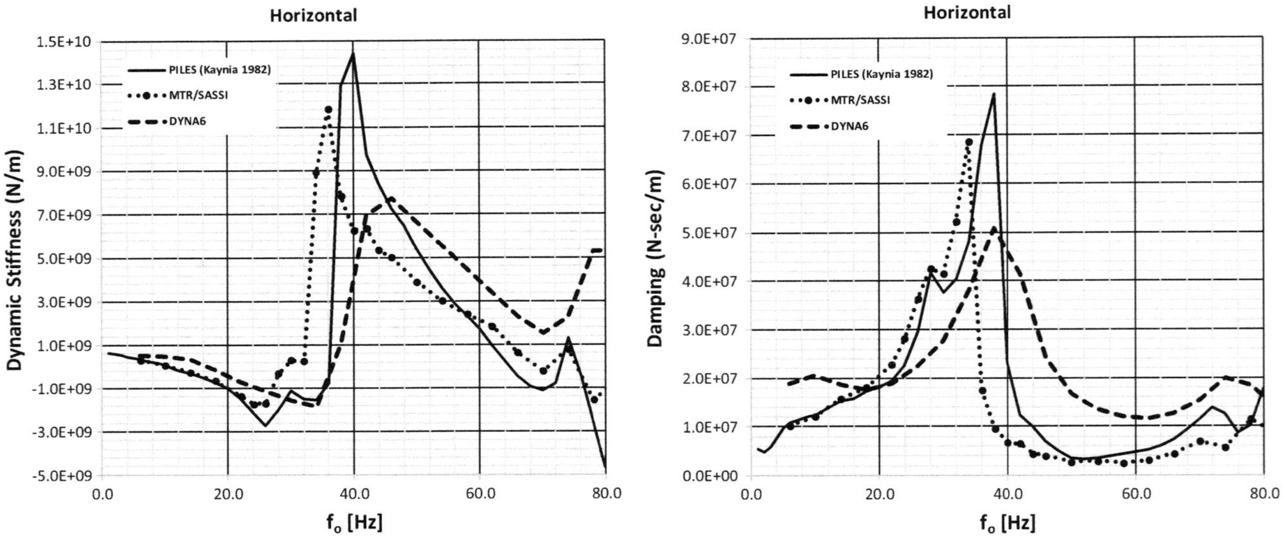

Fig. C-6. Horizontal dynamic stiffness and damping coefficient for the 6 by 6 pile group shown in Fig. C-4

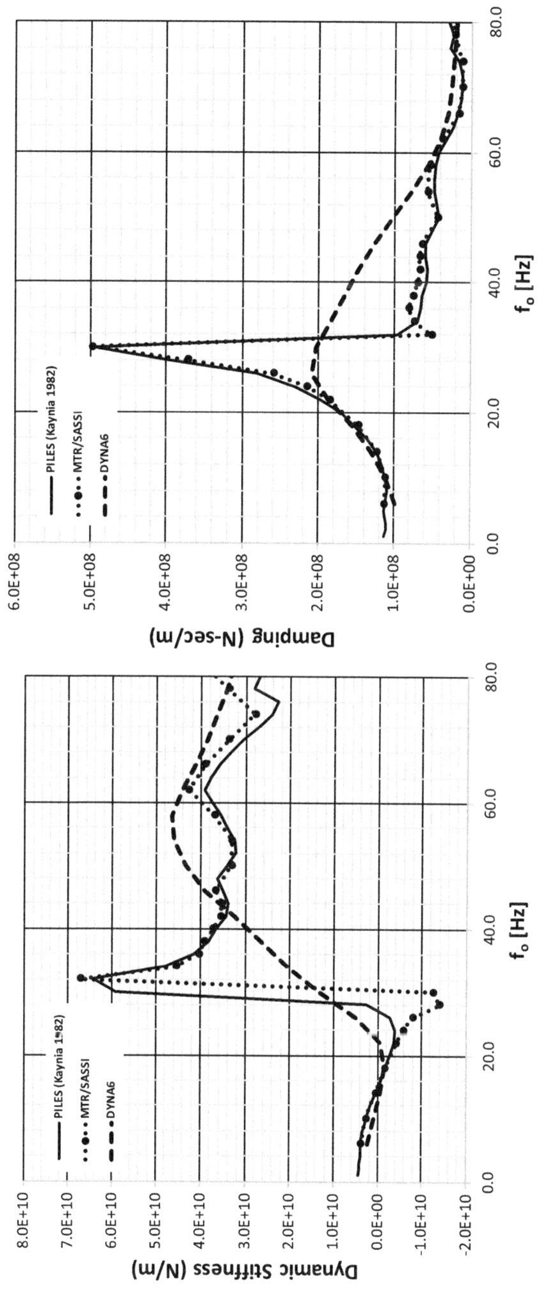

Fig. C-7. Vertical dynamic stiffness and damping coefficient for the 6 by 6 pile group shown in Fig. C-4

Figs. C-6 and C-7 respectively show the dynamic stiffness and damping coefficients for horizontal and vertical vibrations of the pile group, as reported by the previous computer codes.

C.6 NEGATIVE DYNAMIC IMPEDANCE

This example illustrates the solution of the motion equation when the real part (*spring coefficient*) of the dynamic impedance becomes negative. For this purpose, the foundation from Section C.1 is reanalyzed here for a soil profile resulting on negative dynamic stiffness values.

Motion Equation

The generic foundation shown in Fig. 5-2 is used to illustrate the degrees of freedom considered while deriving the motion equation. It must be noticed that the eccentricity between the CG and CR is zero. Furthermore, the foundation is under a vertical harmonic force $F_z(t)$ acting along the CG; therefore the only displacement experienced by the foundation is $uz(t)$. The soil resistance, acting at the CR, is provided by $Pz(t) = K_z u_z(t)$, where Kz is the frequency dependent vertical dynamic impedance. The motion equation is given by Eq. (5-2), detailed as follows:

$$K_z(\Omega)\cdot u_z(t)+m\cdot \ddot{u}_z(t)=F_z(t)$$

or considering a spring damper system for the soil resistance

$$(K'_z(\Omega)+\Omega\cdot i\cdot C_z(\Omega))\cdot u_z(t)+m\cdot \ddot{u}_z(t)=F_z(t)$$

For harmonic excitation, $F_z(\mathrm{t})=F_o e^{i\Omega t}$ the above equations reduce to

$$(K_z(\Omega)-\Omega^2 m)\cdot u_z(\Omega)=F_o$$

or

$$(K'_z(\Omega)+\Omega\cdot i\cdot C_z(\Omega)-\Omega^2 m)\cdot u_z(\Omega)=F_o$$

The displacement is calculated as follows:

$$u_z(\Omega)=\frac{F_o}{(K_z(\Omega)-\Omega^2 m)}$$

or

$$u_z(\Omega)=\frac{F_o}{(K'_z(\Omega)+\Omega\cdot i\cdot C_z(\Omega)-\Omega^2 m)}$$

Foundation Properties

For the sake of simplicity, a circular block foundation with the following dimensions is considered here.

Block foundation radius	$r_f := 1\ m$
Block foundation thickness along z-axis	$t_f := 0.5\ m$
CR eccentricity with respect to CG	$z_c := 0.25\ m$
Embedment depth	$d_f := 0\ m$
Reinforced concrete density	$\rho_c := 2400 \dfrac{kg}{m^3}$
Foundation plus machine mass	$m_o := \pi \cdot r_f^2 \cdot t_f \cdot \rho_c + 2000\ kg = 5770\ kg$

Dynamic Forces Resulting from the Machine

Operating speed of machine	$f := 30\ Hz$
Rotating mass	$m_e := 500\ kg$
Rotor eccentricity	$e_r := 0.0068 \cdot in$
Location of rotor with respect to CG	$z_r := 0.25\ m$
Circular operating frequency of machine	$\Omega_o := 2 \cdot \pi f = 188.496 \dfrac{1}{s}$
Unbalance force	$F_o := m_e \cdot e_r \cdot \Omega_o^2 = 3.068 \cdot kN$

Frequency Dependent Load

The unbalance force calculated varies with machine frequency during the startup and shutdown operations of the machine. Therefore, the frequency dependent load is calculated as follows:

$$F_\Omega(\Omega) := \left(\frac{\Omega}{\Omega_o}\right)^2 \cdot \mathrm{F_o}$$

$$F_\Omega(\Omega_o) = 3.068 \times 10^3\ \mathrm{N}$$

Dynamic Impedances of the Foundation

A realistic layered soil profile is selected for this example to illustrate the frequency dependent behavior of the foundation impedances. The variation of the real and imaginary part of these dynamic impedances is shown in Fig. C-8. As can be seen, both impedances are highly frequency dependent. Furthermore, the real part of the dynamic stiffness (spring coefficient) becomes negative for certain frequencies.

As discussed in Section 5.3.2, negative dynamic springs coefficients can be modeled in standard structural analysis codes using the concept of

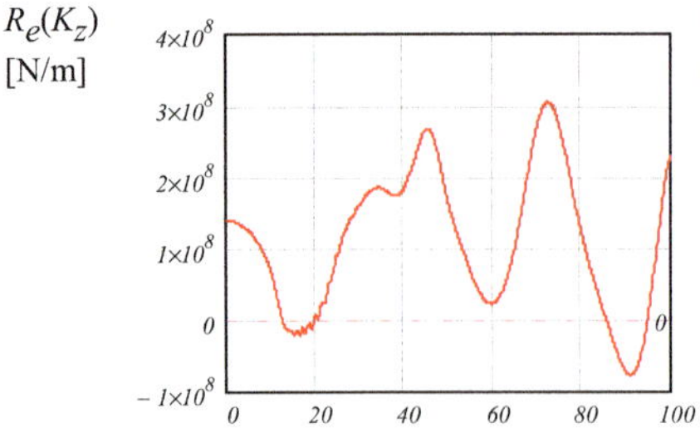

Frequency Hz
(a) Real part of dynamic impedance

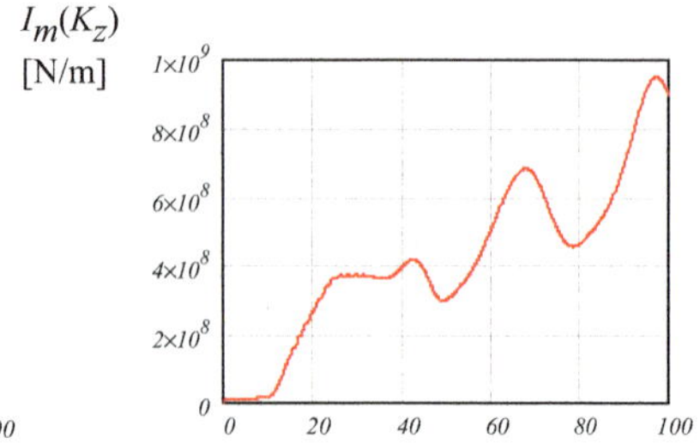

Frequency Hz
(b) Imaginary part of dynamic impedance

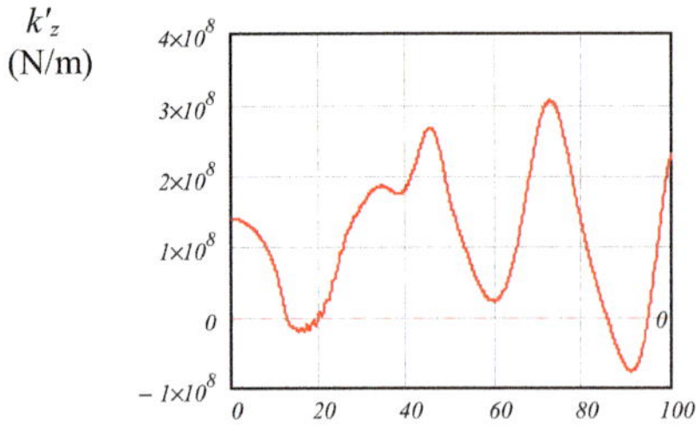

Frequency Hz
(c) Dynamic spring coefficient

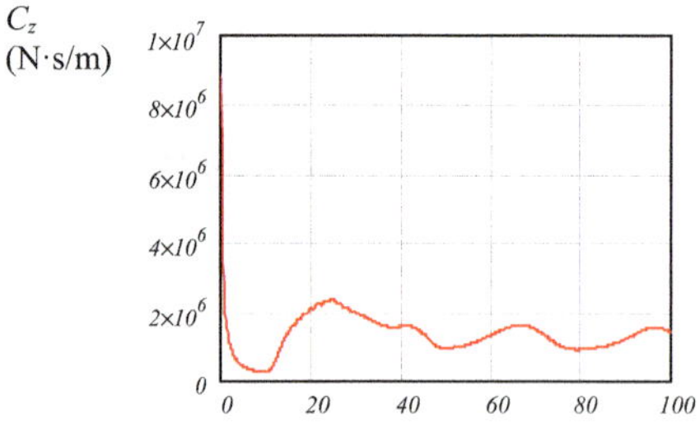

Frequency Hz
(d) Dashpot coefficient

Fig. C-8. Foundation impedances

added mass, which is illustrated next for the frequency dependent spring stiffness. Per Eq. (5-29),

$$K'_z = K_{zo} - \Omega^2 \mathrm{m}_s$$

where K_{zo} is the static stiffness of the foundation ($\Omega = 0$) and m_s is the frequency dependent added mass. Therefore,

$$K_{zo} := K'_z(0) = 1.391 \times 10^8 \cdot \frac{\mathrm{N}}{\mathrm{m}}$$

and the added mass for $K'_z < 0$ is

$$m'_s(\Omega) := if\left(K'_z(\Omega) < 0, \frac{K_{zo} - K'_z(\Omega)}{\Omega^2}, 0 \right)$$

For instance, for

$$\Omega_t := 2 \cdot \pi \cdot 18\ Hz = 113.097 \cdot \frac{\mathrm{rad}}{\mathrm{s}}$$

$$K'_z(\Omega_t) = -1.803 \times 10^7 \frac{\mathrm{kg}}{\mathrm{s}^2}$$

$$m'_s(\Omega_t) = 1.229 \times 10^4 \mathrm{kg}$$

and the corresponding spring coefficient becomes $k'_z = K_{zo}$ for $K'_z < 0$. Therefore,

$$k'_z(\Omega) := if(K'_z(\Omega) < 0, K_{zo}, K'_z(\Omega))$$

$$k'_z(\Omega_t) = 1.391 \times 10^5 \cdot \frac{\text{kN}}{\text{m}}$$

Fig. C-9 shows the variation of the added mass and corresponding spring coefficient using these definitions for m'_s and k'_z. As can be seen, the spring coefficient does not become negative for any frequency.

Solution of the Motion Equation

The inversion of the motion equation can be done directly using complex impedance values or the concept of added mass as illustrated next.

Harmonic displacement calculated using complex impedance:

$$u_z(\Omega) := \frac{F_\Omega(\Omega)}{K_z(\Omega) - \Omega^2 \cdot m_o}$$

Harmonic displacement calculated using real spring-dashpot coefficients and added mass:

$$u'_z(\Omega) := \frac{F_\Omega(\Omega)}{k'_z(\Omega) + \Omega \cdot i \cdot C_z(\Omega) - \Omega^2 \cdot (m_o + m'_s(\Omega))}$$

Displacements of the Center of Mass

The CG displacements at the machine operating frequency Ω_o are calculated next, using the previous equations. Note that the displacement amplitude is the quantity of interest.

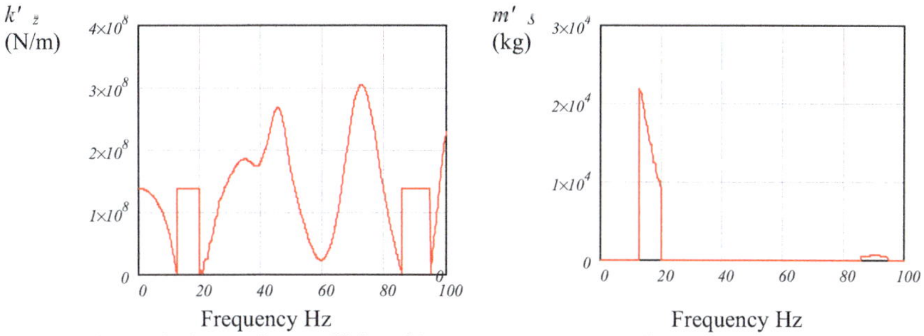

(a) Equivalent spring coefficient k'_z (b) Added mass m'_s

Fig. C-9. Spring and added mass coefficients

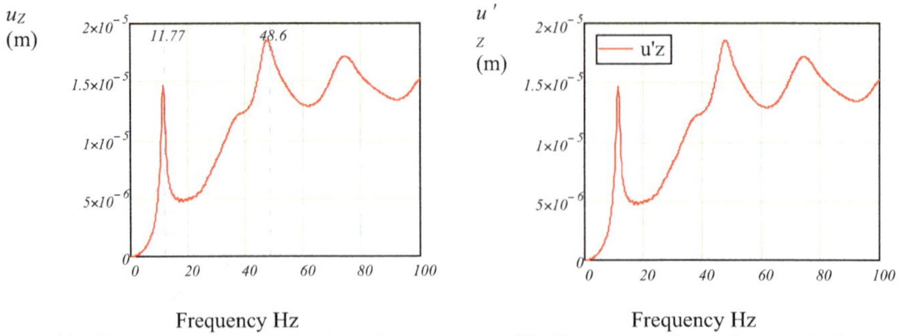

(a) uZ response using complex impedance

(b) uZ response using real spring-dashpot coefficients and added mass

Fig. C-10. Startup and shutdown response of rigid block foundation

Maximum CG displacement using complex impedance:

$$|u_z(\Omega_0)| = 8.196 \times 10^{-6} \text{ m}$$

Maximum CG displacements using real spring-dashpot coefficients and added mass:

$$|u'_z(\Omega_0)| = 8.196 \times 10^{-6} \text{ m}$$

Additionally, the shutdown and startup response of the foundation is shown following. This information is helpful to investigate the foundation response for the whole range of operating frequencies of the machine, and identify critical frequencies (e.g., 11.77 Hz and 48.6 Hz) as shown in Fig. C-10.

As shown in this example, the dynamic response of a machine foundation can be calculated using complex dynamic impedances or real spring-dashpot coefficients and added mass. Both approaches yield identical results. However, the added mass approach is easier to implement using standard structural analysis programs.

REFERENCES

Arya, S. C., O'Neill, M. W., and Pincus, G. (1984). *Design of structures and foundations for vibrating machines*. Gulf Publishing, Houston.

Bowles, J. E. (2007). *Foundation analysis and design*. Knovel, Norwich, NY.

Dobry, R., and Gazetas, G. (1988). "Simple method for dynamic stiffness and damping of floating pile groups." *Géotechnique* 38(4), 557–574.

El Naggar, M. H., and Novak, M. (1995a). "Nonlinear lateral interaction in pile dynamics." *Soil Dyn. Earthquake Eng.* 14(2), 141–157.

El Naggar, M. H., Novak, M., Sheta, M., El-Hifnawy, L., El-Marsafawi, H., and Ramadan, O. (2011). *DYNA6 a computer program for calculation of foundation response to dynamic loads*. Geotechnical Research Centre, University of Western Ontario, London, Ontario.

Gazetas, G. (1991). "Formulas and charts for impedances of surface and embedded foundations." *J. Geotech. Engrg. Div.* 117(9), 1363–1381.

Kaynia, A. M. (1982). "Dynamic stiffness and seismic response of pile groups." *Res. Rep. R82-03*, Dept. of Civil Eng. Massachusetts Institute of Technology, Cambridge, MA.

MTR/SASSI. (2015). "System for analysis of soil-structure interaction," Version 9.4.6.7, MTR & Associates, Lafayette, CA.

Novak, M. (1974). "Effect of soil on structural response to wind and earthquake." *Earthquake Eng. Struct. Dyn.* 3(1), 79–96.

Novak, M. and El-Sharnouby, B. (1983). "Stiffness constants of single piles." *J. Geotech. Engrg. Div.* 109(7), 961–974.

Tabatabaie, M. (2015). "Dynamic impedance of large pile-group foundations." *Transactions*, SMiRT-23. Manchester, UK, August 10–14, Div. V.

Tabatabaie, M. (2017). "On the prediction of experimental results from two pile tests under forced vibrations." *Transactions*, SMiRT-24. BEXCO, Busan, Korea, August 20–25, Div. V.

INDEX

Page numbers followed by *e*, *f*, and *t* indicate equations, figures, and tables, respectively.